CAMBRIDGE LIBRARY COLLECTION

Books of enduring scholarly value

Life Sciences

Until the nineteenth century, the various subjects now known as the life sciences were regarded either as arcane studies which had little impact on ordinary daily life, or as a genteel hobby for the leisured classes. The increasing academic rigour and systematisation brought to the study of botany, zoology and other disciplines, and their adoption in university curricula, are reflected in the books reissued in this series.

Monocotyledons

Agnes Arber (1879–1960) was a prominent British botanist specialising in plant morphology, who focused her research on the monocotyledon group of flowering plants. She was the first female botanist to be elected a Fellow of the Royal Society, in 1946. This volume, first published as part of the Cambridge Botanical Handbooks series in 1925, provides an anatomical and comparative study of the monocotyledon group of plants with an analysis of the methods and objects of studying plant morphology. At the time of publication, comparative anatomy and morphology were the centre of botanical investigation; however there were differences between British and continental biologists concerning the aims of morphological study. In the introduction to this volume Arber reconciled these views by describing a distinction between pure and applied morphology, interpreting the differences in monocotyledonous species in light of this. The book contains an extensive bibliography and 160 figures.

Cambridge University Press has long been a pioneer in the reissuing of out-of-print titles from its own backlist, producing digital reprints of books that are still sought after by scholars and students but could not be reprinted economically using traditional technology. The Cambridge Library Collection extends this activity to a wider range of books which are still of importance to researchers and professionals, either for the source material they contain, or as landmarks in the history of their academic discipline.

Drawing from the world-renowned collections in the Cambridge University Library, and guided by the advice of experts in each subject area, Cambridge University Press is using state-of-the-art scanning machines in its own Printing House to capture the content of each book selected for inclusion. The files are processed to give a consistently clear, crisp image, and the books finished to the high quality standard for which the Press is recognised around the world. The latest print-on-demand technology ensures that the books will remain available indefinitely, and that orders for single or multiple copies can quickly be supplied.

The Cambridge Library Collection will bring back to life books of enduring scholarly value (including out-of-copyright works originally issued by other publishers) across a wide range of disciplines in the humanities and social sciences and in science and technology.

Monocotyledons

A Morphological Study

AGNES ARBER

CAMBRIDGE
UNIVERSITY PRESS

CAMBRIDGE UNIVERSITY PRESS

Cambridge, New York, Melbourne, Madrid, Cape Town, Singapore,
São Paolo, Delhi, Dubai, Tokyo, Mexico City

Published in the United States of America by Cambridge University Press, New York

www.cambridge.org
Information on this title: www.cambridge.org/9781108013208

© in this compilation Cambridge University Press 2010

This edition first published 1925
This digitally printed version 2010

ISBN 978-1-108-01320-8 Paperback

Cambridge Botanical Handbooks

Edited by A. C. SEWARD

MONOCOTYLEDONS

CAMBRIDGE
UNIVERSITY PRESS
LONDON: Fetter Lane

NEW YORK
The Macmillan Co.

BOMBAY, CALCUTTA and
MADRAS
Macmillan and Co., Ltd.

TORONTO
The Macmillan Co. of
Canada, Ltd.

TOKYO
Maruzen-Kabushiki-Kaisha

Tamus communis, L.; Black Bryony (Dioscoreaceae)

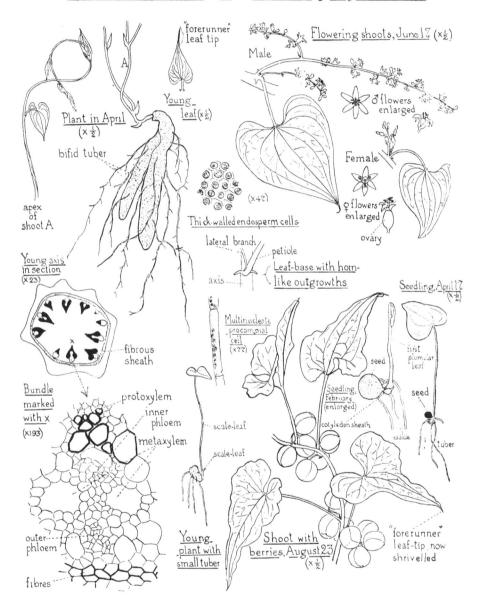

"forerunner" leaf tip

Flowering shoots, June 17 (×½)

Male

♂ flowers enlarged

Young leaf (×½)

Plant in April (×½)

bifid tuber

apex of shoot A

Female

♀ flowers enlarged

ovary

Thick-walled endosperm cells

(×47)

Young axis in section (×23)

lateral branch

petiole

axis

Leaf-base with horn-like outgrowths

fibrous sheath

Seedling, April 17 (×½)

first plumular leaf

seed

Multinucleate procambial cell (×77)

Bundle marked with x (×193)

protoxylem

inner phloem

metaxylem

Seedling, February (enlarged)

seed

cotyledon sheath

radicle

tuber

scale-leaf

outer phloem

scale-leaf

Young plant with small tuber

Shoot with berries, August 23 (×½)

"forerunner" leaf-tip now shrivelled

fibres

MONOCOTYLEDONS

A MORPHOLOGICAL STUDY

BY

AGNES ARBER, M.A., D.Sc.

SOMETIME FELLOW OF NEWNHAM COLLEGE, CAMBRIDGE

WITH A FRONTISPIECE AND
ONE HUNDRED AND SIXTY FIGURES

CAMBRIDGE
AT THE UNIVERSITY PRESS
1925

TO

THE MEMORY OF

ETHEL SARGANT

PREFACE

SOME fifteen years ago, the late Ethel Sargant accepted the invitation of the Editors to write a book on Monocotyledons for this series. It is a matter for the deepest regret that the failure of her health prevented her from using this opportunity of treating a subject, which she had made peculiarly her own, in more comprehensive fashion than was possible within the limits of her published memoirs. Shortly before her death in 1918, when she was beginning to feel that her physical powers had become unequal to any prolonged strain, she suggested that I should undertake the task in her stead. I had hoped that among her manuscripts, which were left in my hands, there might have been some fragments intended for the book, which it would have been possible to utilise; but with the exception of a number of drawings, and a few notes which I have been able to quote, I found nothing which I could incorporate. For the present volume I am, therefore, wholly responsible. By the generosity of the Mistress and Council of Girton College, I have had the use of Miss Sargant's library, and also of her collection of slides; these, in addition to a bibliography of the literature relating to Monocotyledons, which she had begun to compile, have been of the utmost service in connexion with the present study. But this book—or indeed any other piece of work falling to my lot—owes more to Ethel Sargant than can be conveyed by any statement of specific indebtedness. As her assistant, and, afterwards, her colleague in joint investigation, I received from her a training for which no expression of gratitude can be adequate. That my study of Monocotyledons has led me to depart, fundamentally, from the views which she herself advocated, is a result which she would have welcomed: she was keenly alive to the fact that scientific hypotheses have in their nature no pretension to permanence, and that they should be judged by their capacity for bringing to light further generalisations, to which, in turn, they yield their place. To work with Ethel Sargant was to realise the pursuit of science as an unending adventure of the mind: in dedicating this book to her memory, I dedicate it to the very spirit of research.

AGNES ARBER

BALFOUR LABORATORY,
  CAMBRIDGE.
    February 20, 1925.

ACKNOWLEDGEMENTS

For the living plants and the herbarium material, which I have used in the present study, and for help and criticism in the course of my work, I wish to express my thanks to the Director of the Royal Botanic Gardens, Kew; the Director and the Superintendent of the Cambridge Botanic Garden; the Director of the Botanic Gardens, Sydney, N.S.W.; the Director of the National Botanic Gardens, Kirstenbosch, Cape Town; the Superintendent of the Royal Botanic Gardens, Sibpur, Calcutta; and to Mr J. Burtt-Davy; Messrs Charlesworth, of Hayward's Heath; Mr W. R. Dykes; Mr T. A. Dymes; Mr Somerville Hastings; Mr J. Hutchinson; Miss Gulielma Lister; Sir Herbert Maxwell, Bt.; the Hon. Mrs Huia Onslow; Mr J. Ramsbottom; Miss E. R. Saunders; and Miss Winifred Smith.

About one hundred and forty of the illustrations in the present book are either new (initialed A. A.), or derived from my own published papers; in this connexion I have to record my special indebtedness to the Editor of *The Annals of Botany* and to the Clarendon Press; I have also to thank the Editors of other journals in which certain of the figures have appeared— *The Botanical Gazette, The New Phytologist* and *The Journal of the Indian Botanical Society*—as well as the Councils of the Royal Society and the Linnean Society. In the case of those text-figures, numbering about twenty, which are derived from the work of other authors, the reference to the source will be found in each legend, but I must here acknowledge my obligation to Mr J. M. Black, of North Adelaide, and the Royal Society of South Australia, for the block for Fig. clv; Miss M. B. Church and the Torrey Botanical Club, for the use of Figs. xxiv and xxv; Dr T. W. Woodhead and the Editor of *The Naturalist*, for the block for Fig. vi; Prof R. H. Yapp and the Editor of *The New Phytologist*, for that for Fig. xxvi; while Figs. xl, ciii, cxli and cxlii are reproduced, by courtesy of the Essex Field Club, from blocks which illustrated Miss Gulielma Lister's presidential address to that Society in 1920.

Finally, I wish to express my gratitude to Professor A. C. Seward, F.R.S., who has been so kind as to read and criticise the proof-sheets.

A. A.

CONTENTS

LIST OF ILLUSTRATIONS

"Toutefois, que dire d'un savant qui déclare s'en tenir à la production, ou à la bonne disposition *de faits positifs*? S'il ne se plaît qu'à bien élaborer ses matériaux, et qu'à les livrer parfaitement façonnés, pour être un jour employés, il renonce à ce qu'il y a de plus vif, de plus enivrant, et de plus profondément philosophique dans la vie des sciences. C'est ne vouloir jouer que le rôle d'un habile appareilleur; c'est, en manifestant bien peu de confiance en soi, vraiment craindre de se hasarder dans les conceptions de l'architecte."

GEOFFROY SAINT-HILAIRE, 1833.

CHAPTER I

INTRODUCTION: THE PRINCIPLES OF MORPHOLOGY

IN recent years botanical morphology has met with much destructive criticism from those whose interest lies in other branches of the subject. There is apparently a general tendency to treat *form* as non-essential, almost indeed as if it were something superficial and extraneous—an afterthought, like the façade, with its stuccoed pilasters, plastered on to the front of a Victorian villa. But, in reality, the lines and surfaces of the living body are determined by inner necessity, and are more full of significance even than those of a vaulted roof or of a flying buttress. There are some biologists, again, who would condemn morphology because they think that it should be replaced by bio-chemistry and bio-physics. It is true that analysis may in course of time be carried to a point at which it becomes possible to give a complete expression for the form of a living thing in terms of chemical and physical formulae. But that consummation is not yet reached, and meanwhile we at least possess, in the form itself—using the word in its widest sense—the final synthesis of all such partial expressions. Form is, indeed, chemistry and physics made manifest.

In the present book I shall employ the term Morphology to mean the study of plant organs from the standpoint of their form relations, the word "form" being taken to include structural as well as external features. Like mathematics, morphology may be classified under two headings—"pure" and "applied." Pure morphology is concerned with the comparative examination of form, studied in itself and for its own sake; it is directed to the discovery of what Augustin-Pyramus de Candolle, nearly a century ago, called "les lois générales de la symétrie organique." Applied morphology, on the other hand, seeks to use the evidence of form to elucidate evolutionary history. Pure morphology is the older branch, and it has elements of permanence which applied morphology lacks; it is the fundamental and basic aspect of the study. Applied morphology is a secondary development, which sprang into existence as soon as Darwin's work won general acceptance for the doctrine of evolution, and it is a misfortune that its rapid but largely unsound growth soon overshadowed morphology proper. Applied morphology has generally been based upon the assumption that the evolutionary history of the vegetable kingdom could be visualised as a tree, of which the terminal twigs represent members of living groups. On this assumption, the search for synthetic ancestral types was widely undertaken, and the attempt to trace phylogenies by means of morphological evidence was made on all hands. But modern work in palaeobotany has revealed a surprising lack of synthetic types, and every fresh step in research seems to carry the origin of the great groups farther and farther back in time; the image of the tree thus constantly recedes from our vision, while its place is taken by a series of more or less parallel life-lines, whose origin is lost in remote antiquity. We have at last come to realise something of the supremely important part which has been played by parallelism of development, and we no longer hold it *necessary* to invoke descent from a common stock to account for resemblances between the plants of to-day. I think, then, that it must be admitted that the applied morphology, which was developed as an offshoot from the Darwinian theory, and which aimed at the discovery of phylogenetic connexions between the larger groups, has proved bankrupt, because it depended on a certain definite conception of evolutionary history, which the progress of research has since invalidated.

We are thus—for the present, if not for ever—compelled to give up the hope of tracing the relations of the larger groups on morphological grounds. But when we turn to the only aspect of applied Morphology which directly concerns us in the present book, and consider the phylogenetic questions which arise when we are dealing with Angiosperms alone, the situation has to be judged afresh. It seems, on analysis, that our attitude towards the applied morphology of Flowering Plants must depend on the answers to the following questions: are we to treat, as indications of real genetic affinity, those resemblances which systematists have recognised by grouping species into genera, and genera into families? Or are we to suppose that these genera and families include an assemblage of unrelated plants, which have arrived by different but converging routes at the same plane of development? We can get no help from palaeobotany in deciding between these two alternatives, for the study of fossil plants has at present thrown no light on the origin of the existing groups of Angiosperms; but it is with geographical botany that the decision seems to rest. For when the areas occupied by the genera belonging to any given family[1], or the species belonging to any given genus, are mapped, the relation of the areas of distribution is, in many cases, impossible to account for, except on the theory that there is some kind of genetic connexion between the species and genera in question. But farther than this we cannot go, for, unlike the genera and species, the families of Angiosperms have, in most cases, a geographical distribution which is too wide and too uniform to serve as a basis for deductions as to their history; the question, indeed, of how the great Angiospermic families, with their distinctive characters, came into existence, still remains an unsolved mystery. We know nothing about the course of the constructive

[1] I am only concerned here with such families and genera as have been recognised by systematists, after critical study, as "natural."

up-grade evolutionary process, which, on this particular line, culminated in the Angiospermic type of plant construction. But we can, on the other hand, dimly discern the features of the *down-grade* process—called by Guppy "differentiation"—in which reduction plays such a conspicuous part, and which leads from the generalised family type, by endless variations on the given theme, to the bewildering variety of specific forms which we recognise to-day. It is of peculiar interest to find that the greatest of the early morphological writers—Goethe in his *Versuch* of 1790, and A.-P. de Candolle in his *Organographie végétale* of 1827—are primarily concerned with the process which results in the "differentiation" of the family type into genera and species, though their language, naturally, is not that of the modern evolutionist. De Candolle points out that, on his view, "chaque famille de plantes...peut être représentée par un état régulier, tantôt visible par les yeux, tantôt concevable par l'intelligence; c'est ce que j'appelle son *type*." And he proceeds to show that the characters of the members of the family result from the modification of this "type primitif," by the fusion, abortion, degeneration and multiplication of parts. It seems to me that applied morphology, even at the present day, still finds its most hopeful field in the comparatively modest task of pursuing that study of the "differentiation" process, which was begun by de Candolle nearly a hundred years ago, on foundations laid still earlier by Goethe.

The applied morphology of Flowering Plants, limited in the sense just indicated, should aim, I think, in the first place at tracing the homologies of organs. When thus occupied, it is, perhaps, scarcely necessary to qualify it by the word "applied," since it grades insensibly into pure morphology. For it must not be forgotten that the idea of homology, though often treated as if it were indissolubly linked with the doctrine of "descent with modification," had, in reality, an independent origin in men's minds, many centuries before biology became dominated by the evolution theory. Aristotle, for instance, seems to have recognised something approaching to homology, in the modern sense, between the arms of man, the forelegs of quadrupeds and the wings of birds. What the theory of evolution did, was to offer a simple *interpretation* of homology, since it explained the identity of type of two homologous organs by assuming that they are the divergent representatives of a common ancestral form. But it is possible to recognise the existence of homologies, without necessarily accepting this particular explanation of their meaning. We may, for instance, hold that there is a homology between the leaves of the various groups of Angiosperms, without committing ourselves to the theory that these groups are descended from a single stock.

The morphologist who deals with Flowering Plants may either be satisfied with the study of form in itself, and with the tracing of the homologies of individual organs; or he may, on the other hand, wish to go farther, and to attempt—within the limits set by the various main divisions of the Angiosperms—to trace phyletic lines. Now, whether he is concerned with the elucidation of homologies, or with the problems of phylogeny, he must first make up his mind on two points—the nature of the evidence from which he will draw his conclusions, and the nature of the proof by which these conclusions are to be tested. It is by no means easy to formulate in set terms the complete procedure to be adopted when one is faced with a problem in Angiospermic morphology, but it seems to me that the ideal method for determining homologies may be said to consist in the examination of the following points, so far as they are applicable, each being studied not only in itself, but comparatively:

1. The external form at maturity.
2. The development (including both apical ontogeny and the stages between the fertilised egg and the adult individual).
3. The anatomy.
4. The comparison of the normal form with variations and abnormalities.
5. The geographical distribution of the type of form in question.
6. The systematic distribution of the type of form in question.

It is to be hoped that some day a seventh source of evidence will be opened to the morphologist who deals with Flowering Plants—that of palaeobotany.

Each of these lines of evidence has its own inherent limitations, as well as its own peculiar value, and unless as many of these sources as can possibly be made available are habitually laid under contribution, the morphological outlook will inevitably be warped. This certainly sounds a mere platitude, but it is necessary to enunciate it, because even our leading morphologists have often shown the most irrational eclecticism in regard to the evidence which they are willing to admit—an eclecticism which has perhaps had more share than any other factor in bringing the subject into disrepute. Velenovský and Čelakovský, for instance, altogether neglect ontogeny, while neither Velenovský nor Goebel make any effective use of structural evidence—indeed Goebel, in his monumental *Organographie*, practically disregards anatomy.

I must confess that I find a difficulty in understanding the attitude of those botanists, who, while willing to admit the possibility of deducing the homologies of an organ from its external form, yet refuse to allow that its anatomy can give any help towards forming a judgment. This position seems to me, on general grounds, entirely untenable. Philosophically, no valid distinction can be drawn between external form and internal structure. It is true that there is a distinction in practice, founded on the fact that internal structure is invisible to the human eye, except with the aid of special instruments and a special technique. But to make a distinction, on this ground, between the value of the evidence to be drawn from external form and internal structure, is to return to the anthropocentric standpoint of the Middle Ages. This position is surprisingly illustrated in the work of Daněk, a pupil of Velenovský, who, while condemning anatomy as valueless in morphological argument, yet lays great stress on nervation—regardless of the fact that the visible nervation of any organ simply represents those particular facts of vascular anatomy which happen to be observable with the naked eye. It is of course true that all anatomical data are by no means of equal worth as morphological indices, but this criticism applies with the same force to those drawn from external form. In either case the evidence must be subjected to stringent analysis; it is only after prolonged study that one arrives at a sense of relative values in any particular field. As an example of the varying worth of anatomical data, I may say that a study of the vascular structure of Monocotyledonous leaves has led me to the conclusion that the following characters are of the utmost significance from the standpoint of homology:

1. The general plan of the primary vascular system.
2. The orientation of the vascular bundles.

But though I attribute great importance to these two points, there are, on the other hand, certain features of leaf-structure to which, as far as my experience goes, little value can be attached. I find, for instance, that the *amount* of the vascular tissue and fibres, and the *number* of bundles, or even of bundle-series, are often of minimal importance; these characteristics fluctuate in related plants, and even in the same species. In an interesting study on petiole structure, Salisbury[1] has expressed certain conclusions bearing on this subject, which seem to me to need some qualification. This author shows, for various petioles, that an increase in the amount of vascular tissue, or of the number of bundles in cases where secondary thickening is unimportant, accompanies increase in size of the leaf. He considers—and in this I should agree with him—that his observations prove that the functions of the leaf have a profound influence both upon the amount and the arrangement of the xylem, but he goes on to draw the theoretical deduction that "whether we be concerned with the petiolar structure of the mature organism or with that of the cotyledons, it is to function, rather than to phylogeny, that we must look for its elucidation." This conclusion appears to me to be altogether too sweeping. It must be borne in mind that Salisbury's results do not reveal any alteration in the vascular plan of the petioles in question, or in the orientation of the bundles. The adaptational changes to which he draws attention are purely numerical and quantitative: but it is on the evidence of *ground-plan* and *orientation*, rather than of *quantity*, that the

[1] Salisbury, E. J. (1913).

morphologist must rely. More recently Church[1] has emphasised, by means of a pithy and picturesque analogy, the leading part which orientation must always play in morphological thought. He reminds us that an ancient building may be altered out of recognition, and used eventually for some purpose quite alien to that for which it was erected, but that, despite these changes, "certain factors in its original plan, as for example its orientation, will remain unaffected for all time."

Many botanists refuse altogether to admit any evidence belonging to our next category—that of abnormalities. This refusal is no new thing, for nearly a century ago, de Candolle spoke of "cette nombreuse classe de faits, connue sous le nom de monstruosités,...qu'on affectait de mépriser pour se dispenser de les étudier." It is indeed a commonplace that we suffer at present from great difficulty and uncertainty in dealing with the evidence derived from such abnormalities, but this is because our knowledge of them is fragmentary, and we still await a systematic treatment of the subject on broad lines. The result of this unfortunate state of things has been that the majority of our morphologists think themselves justified in disregarding teratology altogether—scared from the ground by the irresponsible attempts which some have made to exploit it. This is greatly to be regretted, and no doubt a rich harvest will be reaped in this field when the laws which govern variations from the specific norm yield to discovery. Indeed, modern thought tends to the conclusion that changes which may fairly be called teratological have played an important part in evolution.

The precise degree of value to be attributed to the evidence from geographical distribution is another controversial question, but probably all botanists would agree that no view as to the genetic relations of different types can be sound, unless it takes into account the facts of their distribution; and it is also generally admitted that there must be some connexion between the age of a type and the area which it occupies[2]. When plant geography has been more fully explored from these standpoints, it will no doubt furnish a mine of information to the morphologist; and even now it often throws an invaluable light on our problems.

It is a matter of difficulty, in the present state of our ignorance, to appraise exactly the worth of the conclusions to be drawn from the sixth and last source of evidence—the systematic distribution of a given type of structure. If a certain type recurs repeatedly in plants which show no close affinity with one another, it is obvious that we need a morphological interpretation of a different order from that which may be adequate for a structure which is confined to one small group of plants. But the subject of the

---

[1] Church, A. H. (1919).
[2] Willis, J. C. (1922), and references there given.

systematic distribution of types of structure is yet in its infancy, and the part such evidence should play in morphology has never received the full recognition which it may be able to claim in the future.

Though the unreasoning eclecticism with which individual morphologists have dismissed valuable sources of evidence seems to me peculiarly unfortunate, I feel that there is even more to regret in the artificial limitations which workers in this field have imposed upon themselves in regard to the organs in which they are willing to look for the evidence which they hold to be valid. The tendency seems to be for each morphologist to select some special region of the plant, and to assume that it is "conservative," and that here, but not elsewhere, ancestral traits may be expected to survive; in recent work, bearing specifically upon phyletic problems, the gametophytes, the seedling hypocotyl, the leaf, the node, the peduncle, the tissues produced in response to wound stimulus, etc., have each in turn been treated as the sole repository of the plant's heirlooms. I cannot but think that the standpoint thus revealed is indefensible; it suggests the materialistic idea that "ancestral traits" are something like old family silver, supposed by tradition to be hidden away in an ancient mansion, while the morphologists in question resemble treasure-seekers, some of whom are firmly convinced that the cache is to be found in the cellars, while others are equally confident that it is useless to look for it except in the attics. Personally, I should prefer not to use the expression "ancestral traits," since I believe that it is, in itself, misleading. I would rather speak of phyletic or homological indications —characters, that is to say, which provide a clue, either to actual affinities, or to some evolutionary tendency of the race. Such phyletic indications seem to me to exist *everywhere* in the plant, but while in some regions they are obvious, in others an intensive study is needed before they can be detected. I happen to have been particularly impressed myself by the amount of evidence carried by the leaves, but this is simply because, in the course of the last seven years, I have had occasion to cut sections of those of a large number of Monocotyledons; no doubt any other organ, studied in the same way, would have yielded corresponding results. I have been amazed to find how much these leaves vary in anatomy; it appears that anyone who chose to make a sufficiently detailed comparative examination, would find it possible to identify, not only the family, but even, in many cases, the genus and species, from the transverse section of the leaf alone. It is merely a question of acquiring, by long familiarity, the necessary subtlety of discrimination. Studies on certain families from this point of view have already been made by various botanists, and it is interesting to find that the fixity of generic and even specific anatomical distinctions, recognised, for instance, by Sauvageau in the leaves of the Sea-grasses, has more recently been paralleled for the axis by Gatin, whose studies have been concerned with the peduncles of the Liliaceae[1]. Although to the naked eye the leaves of Monocotyledons look to most of us extremely uniform, a sufficiently trained observer can here distinguish differences which are the counterpart of the anatomical divergences to which I have referred. A year after I had made a note of the possibility, which microscopic work had suggested to me, of identifying Monocotyledons with no aid but that of the transverse section of the leaf, I found in one of J. D. Hooker's letters, written while he was botanising in Morocco with G. Maw—the monographer of the genus *Crocus*—" Maw recognises the bulbs by leaf, however long the tall grass they grow amongst." There seems indeed to be no room for doubt that phyletic indications, external and internal, are carried by the leaves, and I believe that the same is true of any and every region of the plant; there is no valid reason for supposing that one part is more likely to be favoured in this respect than another, though the degree of ease with which phyletic indications can be perceived, may well vary in the case of different organs. In the present book, more attention will be paid to vegetative than to floral characters, not because I regard them as more important, but because they seem to me to have been relatively neglected. Vegetative characters are more subtle and less easily formulated than floral ones, and, in order to detect them, our powers of observation often need to be reinforced by the microscope. Even in the best and most authoritative systematic work, the description of the leaves is, not infrequently, poor and uninforming in comparison with the vivid and exact account of the flowers. The relative difficulty of seizing and defining vegetative characters is partly, no doubt, inherent in their very nature, but it also depends in part on the fact that our eye is less educated to them than to floral characters. Many generations of botanists and gardeners have dismissed them as relatively unimportant, and have concentrated their attention on the flower—the gardener for aesthetic reasons and the botanist because he has been obsessed by the labour-saving but erroneous idea that vegetative characters, in contrast to reproductive, are mainly adaptations to the environment, and are thus of too little permanent value to merit any large share of the morphologist's attention. This belief, which has long held sway, and which was crystallised by the definite expression accorded to it by Darwin, is less well-balanced than the much earlier attitude, which we can trace in the writings of some of the herbalists of the sixteenth century. These men regarded the whole plant as an entity, and never contemplated the divorce of reproductive from vegetative characters—a divorce to which we have become so accustomed that it is an effort to realise its essential artificiality.

If we now leave the question of the nature of the evidence available for the solution of morphological problems, and turn to the way in which this

[1] Gatin, V. C. (1920); see also pp. 89, 210.

evidence is to be utilised, and to the nature of the proof, we are confronted by a fundamental difference between the "logical method" and the actual proceedings of the biological investigator. This difference Ethel Sargant always treated with marked insistence, although, so far as I am aware, no reference to it will be found in her published work. She maintained that there was a material distinction between "proof" in the mathematical and in the morphological sense, and she regarded Bishop Butler's *Analogy* as presenting, in its structure, the closest parallel she could find to her ideal of a morphological argument. It may be assumed that she did not mean to refer so much to the high place which Butler accords to analogy, as to his masterly treatment of probability. But though the use of analogy is as dangerous as it is attractive, Butler's deliberate and restrained method to a remarkable extent justifies its employment, and I am inclined to think that we waste many opportunities in botanical morphology because we carry our caution to the point of ignoring analogy altogether. After all, life is one phenomenon, whether manifested in the animal or vegetable worlds, and I believe that we could often get suggestions that would—by analogy—help towards the solution of our problems, if we paid more attention to what zoologists, and especially palaeontologists, have discovered about evolutionary history.

It is obvious that the results of Butler's reasoning differ radically from the type of demonstration to which the student of science is introduced in elementary mathematics, and to which, for the rest of his life, his conception of "proof" too often remains limited. The feature in which Butler's reasoning shows the closest affinity to a morphological argument, is that he was always seeking for *a balance of probabilities*, rather than a demonstrative proof. He felt that his part was achieved if he could show that any doctrine possessed "a very considerable degree of probability"; he realised that, in the nature of the case, nothing more absolute was possible. As the botanist Boccone wrote in the seventeenth century—"on se doit contenter des raisonnements probables, là où on ne peut avoir des démonstrations."

There are, of course, many thinkers to whom this admitted lack of absoluteness, in the argument from probability, is wholly repellent. The mind that has a natural affinity for the type of reasoning employed in chemistry and physics is often quite impervious to morphological argument. But this is not the fault of morphology; to some extent each subject must always have its own canons and its own mental processes.

We have now reached the point of considering the nature of the results to which we hope "applied" morphology will eventually lead. A recent critic has described the aim of the subject as "sterile phylogeny tracing," but I doubt if the morphologists of to-day will accept this dictum as final. Let us suppose, for the sake of argument, that a series of homologies, or possibly the course of a phyletic line, has been determined with some approach to certainty; this is not, however, an end in itself—it is merely "the prelude to adventure." The morphologist has next to attack the problem on the philosophical side, and to enquire whether his result throws any light on the laws of evolution—whether it furnishes an instance of some generalisation already established; or whether it demands the revision of some existing hypothesis; or whether, possibly, it affords a clue to some principle hitherto unperceived. It is true that, so long as the Darwinian theory held undisputed sway, there was some excuse for regarding the unravelling of pedigrees as the final goal of morphology; for the chief laws of evolution were supposed, in those days, to be adequately known, and the morphologist's business was merely to apply them. But now that biologists have gradually and painfully learned that the Natural Selection hypothesis is not the master key to the mysteries of the organic world, the centre of significance of morphological study has shifted, and we look to our results as representing the raw material whence the laws of evolution may eventually be deduced[1].

[1] Some of the lines of thought initiated in this Chapter are pursued in Chapters IX and X, on the basis of the study of the Monocotyledonous plant which occupies Chapters II—VIII.

CHAPTER II

THE ROOT

IT is commonly said that, whereas the radicle of the Dicotyledonous seedling develops into the tap root of the mature plant, in the case of Monocotyledons the radicle is ephemeral, and is replaced at an early stage by a number of adventitious roots. Although there are many exceptions, this statement is broadly true, and serves to point a general distinction between the two classes. The frequency of subterranean stems in Monocotyledons is no doubt favourable to rooting from the axis, for it is well known that even stems that are normally aerial will often produce roots, if subterranean conditions are imitated. The Arabs, for instance, rejuvenate old Date-palms by surrounding them below the leaf crown with a vessel of earth, which soon induces the development of a crop of roots[1]. It is possible that the rooting of buried axes results directly from lack of light; for Priestley and Ewing[2] have recently drawn attention to the fact that stems grown in darkness are liable to excessive root development, and they make the interesting suggestion that this peculiarity is to be attributed to a certain structural change, which is associated with etiolation. They find that etiolated stems are prone to develop a functional endodermis, in species where such a layer is lacking in the stem grown under normal conditions. They suppose that this endodermis checks the outward passage of sap, and hence leads to an accumulation of nutrient material, which determines meristematic activity in the tissues immediately within the endodermis, and thus initiates the production of lateral roots. Such stems thus approximate in their behaviour to roots, in which it is the pericycle, lying within the endodermis, that normally produces the rootlet-forming meristem (*pc. mer.*, fig. viii, I, p. 20).

It is at the nodes of the axis that adventitious roots commonly develop; they are sometimes arranged there with great regularity; for example, in *Calla palustris*, L., where they form a ring (fig. xxxvi, 2A, p. 60). The fact that in many Monocotyledons only a single leaf, completely ensheathing the stem, arises at each node, and that, from this node, very commonly one or more roots develop, brings the roots and individual leaves into particularly intimate connexion in plants belonging to this class. It is probable that Chauveaud's "phyllorhize theory[3]" owed its origin to the striking

---

[1] Lindinger, L. (1908).   [2] Priestley, J. H. and Ewing, J. (1923).
[3] Chauveaud, G. (1914).

examples of this connexion sometimes met with among Monocotyledons. According to this hypothesis, the plant is built up of units distinguished as "phyllorhizes," each of which consists of a leaf and its associated root. Such a view is not difficult to maintain for the seedling of *Cordyline australis*, Hook., which Chauveaud uses as an illustration, since here we have a small number of leaves, each with its corresponding root; but the position becomes very different when the attempt is made to extend this interpretation beyond a few such special cases. Unless the axis was primitively a subterranean organ—which is scarcely probable—nodal rooting, and the consequent association of roots with particular leaves, must be regarded as secondary developments; and such a view is incompatible with the phyllorhize theory.

The adventitious roots of Monocotyledons are often annual organs; the crops of roots produced each spring by Crocus corms and Hyacinth bulbs are familiar examples. But in other members of the class the roots have a long life. There are a number of large Monocotyledons, whose axes are more or less arboreal, which develop very extensive and permanent root systems; *Agave, Aloe, Dasylirion* and *Furcraea* come under this heading. The roots are long-lived and are increased by the annual development of a fresh crop outside the older ones—each set being thicker than those of the preceding year. The underground system increases in complexity by the branching of the roots at a considerable depth in the soil. The tree Monocotyledons, when grown in Northern Europe, often fail to show this striking development of the root system, since, except under good cultivation, the roots are apt to die off annually[1].

Most of the peculiarities of the Monocotyledonous root can be correlated with one of two anatomical features, which are probably themselves connected—the lack of normal secondary thickening, and the absence of a deep-seated periderm. It seems, for instance, not unlikely that there is some relation between the absence of secondary thickening and the polyarch condition which is so frequent in Monocotyledonous roots. Fig. i illustrates the structure of a mature root of the Water-soldier, *Stratiotes aloides*, L., with its numerous xylem and phloem groups. Fig. i, 1A, shows the small central cylinder surrounded by the wide lacunate cortex characteristic of water plants. The stele, which is drawn in greater detail in Fig. i, 1B, has a well-marked endodermis (*end.*). In the root figured, there are eight protoxylem groups (*px.*), alternating with eight phloem groups, each consisting of one or more sieve-tubes (*s.t.*) with their accompanying companion-cells. It is a question whether there can be said to be any pericycle outside the phloem groups, as each sieve-tube is separated from the endodermis by one element only, which appears to be a companion-cell. A metaxylem

---

[1] Lindinger, L. (1908).

vessel (*mx.*) occurs on the same radius as each protoxylem tracheid, and there appears to be one extra vessel, on the same radius as a sieve-tube; as usual in water plants, the lignification is extremely slight.

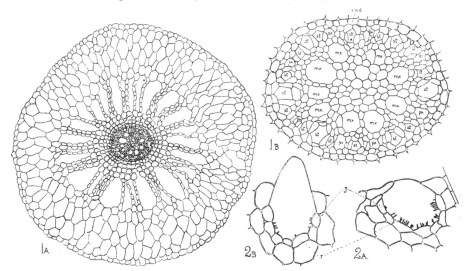

Fig. i. *Stratiotes aloides*, L. Figs. 1 A and B, transverse section of mature root. Fig. 1 A (× 31); Fig. 1 B, stele more highly magnified (× 129); *mx.*, metaxylem; *px.*, protoxylem; *s.t.*, sieve-tube; *end.*, endodermis. Figs. 2 A and B, young root-hair cells from piliferous layer of fresh root, cut in July, to show the rods, *r.*, of colourless, highly refractive material, which rise into the cell cavity of the root-hair from the basal walls; *J*, jacket cells. [A.A.]

In other roots, polyarchy may be carried much farther than in *Stratiotes*; fig. ii, 6, p. 14, for instance, shows a root of *Dracaena* which has 24 protoxylem groups. The thin tertiary root of *Danae racemosa*, (L.) Mönch., drawn in fig. ii, 4, has 10 protoxylem groups, but I have counted 35 in a thick, primary, adventitious root of the same species, and, in other genera, much higher numbers prevail. Many of the smaller Monocotyledonous roots, on the other hand, have few xylem groups; fig. ii, 2, shows a small tetrarch root of *Arum italicum*, Mill., while fig. xiii, 8, p. 27, represents a diarch root-stele from the tuber of *Orchis maculata*, L.

There is great variety among Monocotyledonous roots in the structure of the central region. In *Stratiotes* there is a thin-walled pith, while in the radicle of *Ruscus Hypophyllum*, L., drawn in fig. ii, 3, the outer part of the pith is sclerised. In the tertiary root of *Danae racemosa* shown in fig. ii, 4, the pith is entirely sclerised, but in a secondary and a primary root, which I cut, I found that the pith was thin-walled at the centre, and sclerised at the margins only. Some roots, on the other hand, are pithless; the roots of *Butomus umbellatus*, L., and of *Arum italicum*, Mill., shown in figs. ii, 1 and 2, each have a single metaxylem vessel at the core. In certain Musaceae, Bambuseae, etc., an anomalous-looking structure is produced by

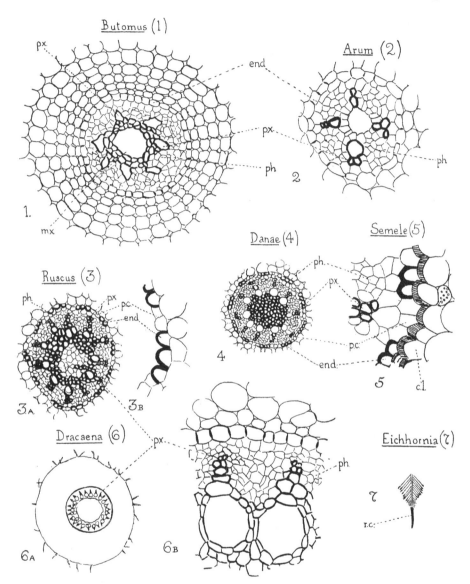

Fig. ii. Root anatomy; throughout, *end.*, endodermis; *p.c.*, passage cells; *px.*, protoxylem; *mx.*, metaxylem; *ph.*, phloem. Fig. 1, *Butomus umbellatus*, L., stele and inner cortex from transverse section of root (× 193). Fig. 2, *Arum italicum*, Mill., stele from transverse section of small root (× 193). Figs. 3 A and B, *Ruscus Hypophyllum*, L.; Fig. 3 A, stele from transverse section of radicle (× 77); Fig. 3 B, endodermis on larger scale to show thin patches and horse-shoe thickening (× 193). Fig. 4, *Danae racemosa* (L.) Mönch., stele from transverse section of thin tertiary root; pith sclerised (× 77). Fig. 5, *Semele androgyna* (L.) Kunth, small bit of margin of stele (× 318) to show endodermis with thin-walled patches, and cortical layer (*c.l.*) next to endodermis, with thickened striated inner walls. Figs. 6 A and B, *Dracaena* sp. ("Duchess of York"); Fig. 6 A, transverse section of root (× 14); Fig. 6 B, margin of stele with two protoxylem groups (× 193). Fig. 7, *Eichhornia speciosa*, Kunth, tip of root with branches (simplified); *r.c.*, root-cap (× ½). [A.A.]

the occurrence of numerous scattered xylem and phloem groups, within the normal ring of alternating patches of xylem and phloem[1].

The piliferous layer of Monocotyledonous roots has an excellent chance of producing interesting developments, since it is neither liable, like that of Dicotyledons, to be crushed out of existence by secondary thickening, nor to be shelled off, with the rest of the outer tissues, by a deep-seated periderm. Fig. iii, 1, shows the piliferous layer, *p.l.*, of the root of *Stratiotes*, from a longitudinal section, cut close to the tip of the root. The piliferous

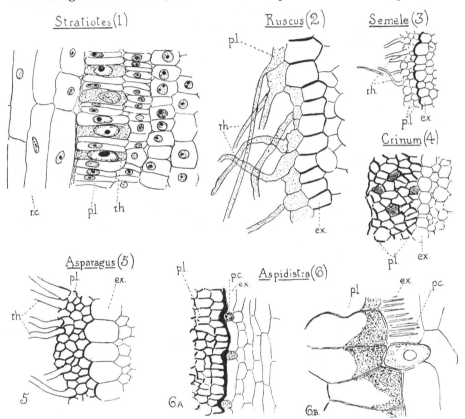

Fig. iii. The piliferous layer; throughout, *p.l.*, piliferous layer; *r.h.*, root-hair; *ex.*, exodermis. Fig. 1, *Stratiotes aloides*, L., longitudinal section near root-tip, showing inner region of root-cap (*r.c.*) and piliferous layer, in which the cells destined to form root-hairs are already distinguishable (× 318). Fig. 2, *Ruscus Hypophyllum*, L., part of margin of transverse section of radicle (× 193). Fig. 3, *Semele androgyna* (L.) Kunth, margin of transverse section of radicle showing two-layered piliferous layer (× 77). Fig. 4, *Crinum Powellii*, Hort., margin of transverse section of root, to show multiple piliferous layer (× 318). Fig. 5, *Asparagus Sprengeri*, Reg., margin of transverse section of small root to show multiple piliferous layer (× 193). Figs. 6 A and B, *Aspidistra elatior*, Blume; Fig. 6 A, margin of longitudinal section of root, showing multiple piliferous layer, and exodermis with passage cells, *p.c.* (× 77); Fig. 6 B, a few of innermost cells of piliferous layer, with adjacent cells of exodermis and cortex, shown in greater detail (× 318). [A.A.]

---

[1] Ross, H. (1883).

layer is at a rudimentary stage, for it is still covered by the root-cap (*r.c.*), but the cells which are destined to form root-hairs (*r.h.*) are already distinguishable by their large nuclei and rich contents. Their later development offers one or two curious features. Before they protrude at all to the exterior, they enlarge so much that they displace the surrounding layers, and finally, by the division of adjacent cells, the base of the root-hair comes to be enclosed in a cup-like jacket of small cells (*J*, fig. i, 2 A and B, p. 13). Another peculiarity shown in these drawings is the presence of delicate, simple or branched rods (*r.*) of a highly refractive substance, which arise from the basal wall of the hair-cell and project slightly into its interior[1].

The layer immediately below the piliferous layer is known as the *exodermis* (*ex.* in figs. iii, 2, 3, 4, 5, 6 A and B, p. 15); it is commonly sclerised, but its continuity as a barrier may be broken by passage cells (*p.c.*, figs. iii, 6 A and B).

The piliferous layer may remain simple throughout the life of the plant, or it may, on the other hand, form a tissue from two to many layers in thickness[2]. Examples of this multiple piliferous layer, or velamen, are shown in figs. iii, 3, 4, 5, 6, which represent sections of the root-surface of *Semele androgyna* (L.) Kunth, *Crinum Powellii*, Hort., *Asparagus Sprengeri*, Reg., and *Aspidistra elatior*, Blume. The cells of the velamen are dead and empty, and sometimes delicately reticulate (e.g. *Crinum Powellii*, fig. iii, 4). A noteworthy feature of the piliferous tissue of *Aspidistra* is the accumulation of stainable material on the inner walls of the innermost layer of the velamen; the granules of which it consists are heaped up, in noticeable fashion, opposite the passage cells (figs. iii, 6 A and B). All the roots illustrated in fig. iii are subterranean, but the most familiar example of a multiple piliferous layer is the velamen which clothes the aerial roots of many Orchids and Aroids. In epiphytic Orchids, the velamen may consist of a single cell-layer, or of a few layers[3], or it may form a many-layered tissue. The recent researches of C. E. Moss[4] have confirmed and emphasised the fact that, both in the Orchidaceae and elsewhere, velamen may be abundantly present in earth roots; the text-book view, that the velamen is a special adaptation to epiphytic life[5], thus demands considerable revision.

Like the piliferous layer, the root-cortex of Monocotyledons often has a more prolonged life than that of Dicotyledons, and consequently it more often shows noteworthy changes at maturity. One of these changes is a curious transverse wrinkling of the root-surface, involving the outer cortex, and associated with a shortening of the root; this wrinkling, though more frequent in Monocotyledons, is not confined to them. Sometimes there is dimorphism

---

[1] Kroemer, K. (1903), Pl. I, fig. 19.   [2] Kroemer, K. (1903); Goebel, K. (1922[2]).
[3] Curtis, K. M. (1917).   [4] Moss, C. E. (1923); see also Holm, T. (1904).
[5] E.g., Haberlandt, G. (1914).

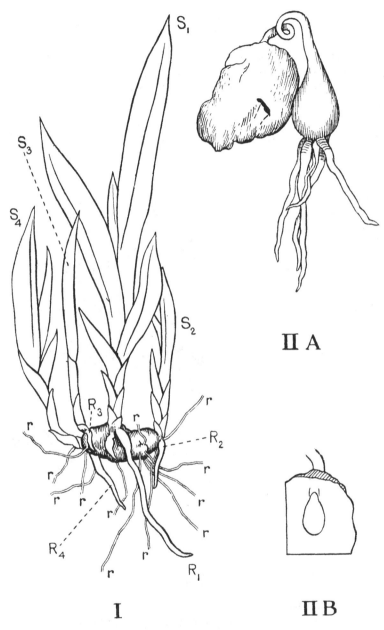

Fig. iv (drawings by Ethel Sargant made at La Mortola in the autumn of 1911).
I, *Antholyza aethiopica*, L., habit drawing of plant, one-third nat. size, showing
S_1—S_4, leafy shoots; *r, r, r,* some of the numerous slender spreading roots;
R_1—R_4, stout roots, probably contractile. II, *Crinum Powellii*, Hort., A, seedling
(slightly reduced), B, apex of cotyledon in endosperm (nat. size).

among the roots, which are not all contractile. This feature is shown in the accompanying illustration of *Antholyza aethiopica*, L. (fig. iv, p. 17), where there are slender roots (*r*) and also thick roots (R_1—R_4), which would, most likely, contract and become wrinkled at a later stage. A similar difference is shown in fig. v, I A, which represents non-flowering bulbs of the Wild Hyacinth, *Scilla festalis*, Salisb.[1], in their immature elongated phase, in which there is one thick, contractile root (*c.r.*) and a number of thin, non-wrinkled roots (*n.c.r.*). Fig. vi, which is taken from Woodhead's interesting study of the wild life of this plant, shows the depth ultimately reached by the mature

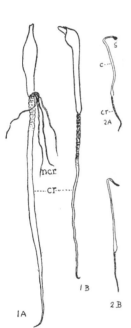

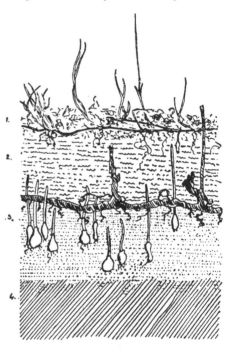

Fig. v. *Scilla* bulbs. I A and B, *Scilla festalis*, Salisb., elongated bulbs in June (×½); *c.r.*, contractile roots; *n.c.r.*, non-contractile roots. 2 A and B, *Scilla autumnalis*, L., seedlings nearly three months from sowing, showing contractile roots and bulbous swelling at base of cotyledon, *c.*, of which the apex in 2 A is still enclosed in seed-coats (×½). [A.A.]

Fig. vi. Section of soil in an area of Birks Wood, near Huddersfield, densely populated by the Wild Hyacinth, *Scilla festalis*, Salisb.; bulbs deeply buried. 1, 3—4 inches of leaf mould; 2, 4—6 inches of peaty humus; 3, slightly sandy, firm loam; 4, clayey loam and stiff clay. Rhizomes of *Holcus mollis*, L., in top layer; rhizomes of Bracken, *Pteridium aquilinum*, Kuhn, at junction between humus and loam. [Woodhead, T. W. (1904).]

bulbs, in an area in which they are present in large numbers. In this locality, the bulbs of the Wild Hyacinth occupied a layer of loam, next below the humus; rhizomes of the Bracken occurred at the junction between humus and loam.

[1] Woodhead, T. W. (1904).

It seems as though plants which perennate by means of subterranean organs, are characterised, at maturity, by a certain definitive depth-position in the soil, which is reached, either by downward growth of the stem itself, or by the pull of those contractile roots which we are here considering. A good example of a plant in whose life-history contractile roots play an important part, is *Phaedranassa chloracra,* Herb. (Amaryllidaceae), which was studied by Rimbach in the tropical Andes (fig. vii)[1]. In the mature stage, the apex of the bulb is 15—30 cms. deep in the earth, but if seeds are sown at such a depth, the seedlings die. If the seed is sown at the surface, the radicle (fig. vii, 1) may attain a length of 95 mm. and may then, in the course of two or three months, contract to 73 mm.; if the apex remains fixed, this means a lowering of the bulb through a distance of 22 mm. A new root is produced every 1— 1½ months, so that, by the end of a year 8— 12 roots are present, and there is always at

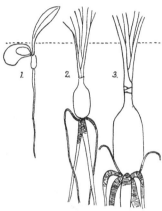

Fig. vii. *Phaedranassa chloracra,* Herb. 1, seedling; 2, plant 6 months old; 3, plant 12 months old. The adaxial parts of the older roots in 2 and 3 are dragged downwards by the pull of the thicker young roots; dotted line indicates surface of earth (× ½). [Rimbach, A. (1895[2]).]

least one root in a state of strong contraction. Each root is longer and thicker than the last, and has a greater power of shortening. The contractile region, which is about 5 mm. long in the radicle, may reach 50—60 mm. in the strong roots of older examples. The maximum rate of contraction is 0·5 mm. in 24 hours. When a plant of *Phaedranassa* gets to its appropriate depth, roots 15—30 cms. long are produced, which have little capacity for contraction.

Both the fact that the roots of Monocotyledons very commonly pass through a contractile phase, and also the fact that the decrease in length is accompanied by a transverse wrinkling of the superficial tissues, are easily demonstrable. But the attempt to determine the exact mechanism of this contraction, and the part played by the different tissues in the process, is fraught with difficulty. Fig. viii, p. 20, shows the vertical rhizome of *Hypoxis setosa,* Baker, which bears long roots, irregular in thickness, whose proximal part is wrinkled and presumably contractile. In order to try to understand how the contraction is brought about, I compared serial sections of the slender smooth zone of a root, 12—15 cms. from the tip, i.e. 17—20 cms. from the base (fig. viii, K), with sections of the wrinkled part of the same root, 2—4 cms. from the base (fig. viii, L). These sections show that it is the

---

[1] Rimbach, A. (1895[2]); see also Rimbach, A. (1895[1]), (1896[1]), (1896[2]), (1897[1]), (1897[2]), (1897[3]), (1897[4]), (1898[1]), (1898[2]), (1900), (1902).

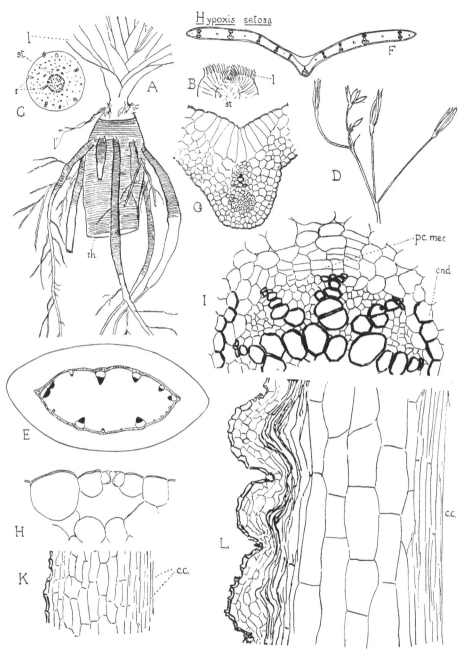

Fig. viii. *Hypoxis setosa*, Baker. Plant from Stellenbosch, S. Africa, grown at Cambridge Botanic Garden, drawn October 5. A, plant ($\times \frac{1}{2}$); *l.*, leaves cut off short; *l'.*, remains of old leaves; persistent roots borne on thick vertical rhizome (*rh.*), here cut off at the base; *i.*, one of several lateral inflorescence-axes. B, top of axis cut in half to show sunk growing point; from the stele, *st.*, numerous leaf-traces are seen passing to the crowded leaf-bases, *l.*, ($\times \frac{1}{2}$). C, axis cut across ($\times \frac{1}{2}$) to show stele, *st.*, roots, *r.*, and mucilage-cavities and bundles in the cortex. D, inflorescence ($\times \frac{1}{2}$). E, transverse section of peduncle ($\times 77$). F, transverse section near apex of leaf ($\times 14$). G, transverse section of midrib of leaf ($\times 77$). H, transverse section of stomate from upper surface of leaf ($\times 318$). I, transverse section of part of stele of slender smooth zone of root, to show pericyclic divisions, *pc. mer.*, giving rise to lateral root; *end.*, cuticularised endodermis interrupted opposite origin of lateral root ($\times 193$). K, longitudinal section of cortex of slender smooth zone of root, 12—15 cms. from tip (17—20 cms. from base); *c.c.* position of central cylinder ($\times 47$). L, longitudinal section of cortex of wrinkled zone of root, 2—4 cms. from base ($\times 47$). [A.A.]

outer tissues which have wrinkled, while the central cylinder and inner cortex are unaffected. But I am wholly unable to understand how the wrinkling comes about, for it seems to me that the elements of the non-wrinkled inner cortex have increased proportionately more in length than have those of the sclerised outermost layer of the root, and yet, instead of being disrupted, the whole of this outermost layer is wrinkled—that is to say, it behaves as if it had increased more in length than the inner tissues. The explanation given by Rimbach[1] is that the shortening is due to a change of form of the inner cortical cells, which he describes as increasing in radial and tangential measurements, while suffering a great decrease in length; a similar view is taken by Woodhead[2]. But the work of the most recent writer on the subject, M. B. Church[3], has again emphasised the baffling nature of the problem, and my attempt to solve it for the case of *Hypoxis* has left me completely puzzled, for I can detect no such shortening as that postulated by Rimbach.

Among Dicotyledons, the formation of root-tubers is generally due to a large parenchymatous development of the secondary vascular tissues. In Monocotyledons, on the other hand, such a process is precluded by the lack of secondary thickening, and the root-tubers are generally found to owe their swelling to a hypertrophy of the cortex, whose long-lived character lends itself to such modifications. *Roscoea cautleoides*, Gagnep. (Zingiberaceae), is an example in which roots—which owe their tuberousness to an exaggeration of the cortex—form the chief perennating organs. Fig. ix, 1 A, p. 22, shows the appearance of the aerial part of the plant at flowering time (May), and also the large bunch of roots, *t.r.*, underground. Fig. 2 A is a drawing of the same plant made in the following December ; at this date the aerial part had all disappeared, leaving only a bud, *b.*, for next year. The roots shown in fig. 1 A are represented by some dead and shrivelled remains (*t.r.*), while a fresh crop (*t.r'.*) has arisen at a slightly higher level. Figs. 2 B—D illustrate the structure of the root marked with an *X* in fig. 2 A. It will be seen that the small stele is surrounded by the wide cortex, which is packed with starch grains.

Triglochin procera, R. Br., is an example of a slightly different type, in which the majority of the roots are slender, while some swell locally into tubers, either at the root apex or farther back. The structure of the tuberous root, t_1, is illustrated in figs. x, B—F, p. 23. Figs. x, E and F, are from sections of the root on the proximal side of the tuber; there is a small tetrarch stele, in which the endodermis and conjunctive are largely sclerised. In figs. x, B and C, on the other hand, we have sections of the tuber itself; there is a slight increase in the size of the stele, as compared with that of the slender part of the root, and a fifth protoxylem group has appeared, but the change

[1] Rimbach, A. (1897[2]).   [2] Woodhead, T. W. (1904).   [3] Church, M. B. (1919).

from the slender to the tuberous form has been brought about, almost entirely, by increase in the cortical parenchyma. It is interesting to note that in the tuber—as compared with the part of the root which forms its

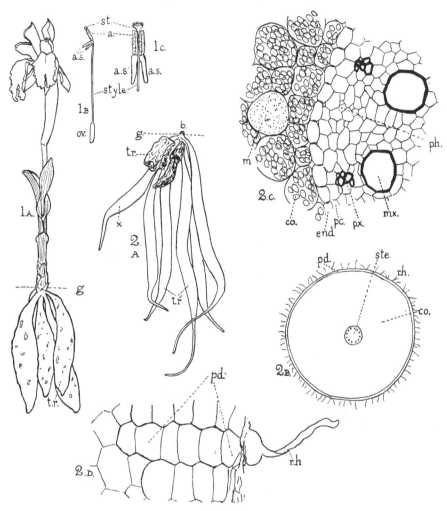

Fig. ix. *Roscoea cautleoides*, Gagnep. (Zingiberaceae). Fig. 1 A, habit drawing, May 26, 1919; *t.r.*, tuberous roots of 1918; *g.*, ground-level; two flowers shown. Fig. 1 B, side view of gynaeceum and androecium (perianth, to which single anther is attached, not shown); *a.s.*, anther spurs; *a.*, anther; *st.*, stigma. Fig. 1 C, front view of stamen and upper part of style and stigma, on a larger scale, to illustrate lever-mechanism. Fig. 2 A, drawing of same plant on December 29, 1919, on a slightly smaller scale; *b.*, bud for next year; 1918 roots, *t.r.*, now dead; *t.r'.*, 1919 roots. Fig. 2 B, transverse section of root marked with an *X* in Fig. 2 A (× 14) to show relation of tiny stele (*ste.*) to wide storage cortex, *co.*; *pd.*, superficial periderm, consisting of about 5 cell-layers; *r.h.*, root-hairs; the 12 black dots in the stele represent 12 xylem groups. Fig. 2 C, margin of stele from fig. 2 B (× 193); *px.*, protoxylem; *mx.*, metaxylem, consisting of one xylem element to each group of protoxylem; *ph.*, phloem; *end.*, endodermis; *pc.*, pericycle; *co.*, cortical cells laden with large starch grains; *m.*, ?mucilage cell. Fig. 2 D, small part of periphery of section drawn in fig. 2 B, to show superficial periderm, *pd.*, and root-hair, *r.h.* (× 193). [A.A.]

stalk—there is a loss of sclerosis in the endodermis and in the conjunctive of the stele.

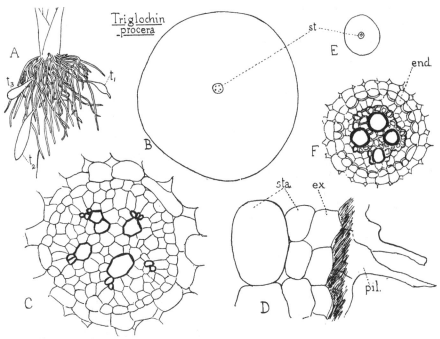

Fig. x. *Triglochin procera*, R. Br., Glenfield, George River, near Sydney, September. A, base of plant (× about ½) with normal roots and three tuberous roots, t_1, t_2, t_3. B, transverse section of root t_1 in A, through its tuberous region (× 14); *st.*, stele. C, transverse section of stele shown in B (× 193). D, margin of section shown in B; *sta.*, cells packed with starch; *ex.*, empty exodermis; *pil.*, remains of piliferous layer (× 193). E, transverse section of root marked t_1 in A, on proximal side of tuber; middle cortex lacunate (× 14). F, stele of E (× 193); *end.*, endodermis. [A.A.]

Figs. xi, A—E, p. 24, represent the root of *Hemerocallis fulva*, L., which, like that of *Triglochin procera*, is locally tuberous, but which differs from the latter in showing a considerable extension of the stele in the tuberous region. Figs. xi, B and C, indicate the contrast in the size of the whole root, and of the stele, in the slender and tuberous parts of the root t_1 in fig. xi, A, while in figs. D and E the same contrast, for the stele alone, is shown on a larger scale. There has been no change in the number of protoxylem groups, which remains seventeen; the increase in size of the stele is chiefly due to expansion of the pith, while the metaxylem elements are also larger.

In *Asphodelus ramosus*, L. (figs. xi, F and G), the stelar tissue plays a still more important part in the root-tuber. Sections were cut from the root marked t_1 and it was found that, whereas the slender distal region had a relatively small stele in which the pith was partially sclerised, the thick part of the tuber, on the other hand had a large stele filled almost entirely by the parenchymatous non-sclerised pith, the xylem and phloem occupying a narrow marginal ring (fig. xi, G).

The morphological nature of the tubers which develop at the apical ends of cylindrical roots in *Dioscorea discolor*, Hort., and other members of this genus, has been the subject of much controversy. But it seems probable that they are, in reality, roots. Their tuberous character is due to hypertrophy of the stelar tissue; as the tuber expands, the bundles become scattered and lose their typically root-like arrangement[1].

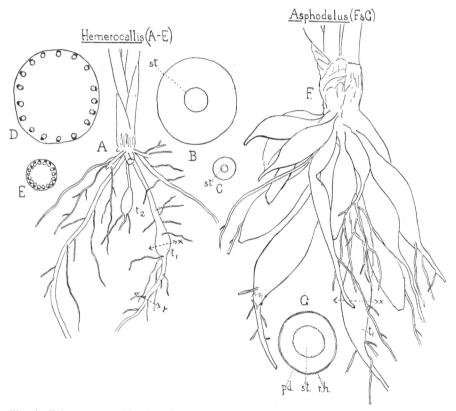

Fig. xi. Tuberous roots with enlarged steles. A—E, *Hemerocallis fulva*, L. A, base of plant in May, with two tuberous roots t_1 and t_2 (× ½). B and C, transverse sections of root t_1 in A, cut at X and Y respectively (× about 3½); *st.*, stele. D and E, stele of B and C (× 14), only xylem indicated. F and G, *Asphodelus ramosus*, L.; F, base of plant with numerous tuberous roots, October 29 (× ½). G, transverse section of root t_1 (nat. size) to show persistent root-hairs, *r.h.*, periderm, *pd.*, stele, *st.*, with large pith. [A.A.]

A very general difference between Monocotyledonous and Dicotyledonous roots lies in the fact that in Monocotyledons the periderm is commonly superficial, while in Dicotyledons it is, as a rule, pericyclic. The interesting work on the endodermis, which we owe to Priestley and his colleagues, throws some light on this distinction. According to the views of this school, the endodermis in many plants in its "secondary" stage is

[1] Lindinger, L. (1907).

rendered impermeable to water and to solutes by the presence of a suberin lamella on the inner surface of the cell-wall. When the walls of all the endodermal cells suffer this change, a cork layer commonly arises in the pericycle—the necessary meristematic activity being stimulated by the accumulation of sap within the endodermal barrier[1]. If, on the other hand, some of the cells fail to pass over into the "secondary" condition, and remain as "passage cells" (*p.c.*, figs. ii, 3, 4, 5, p. 14), the endodermis, owing to the breaches thus left, is an ineffectual barrier, and no sap accumulates within it. But the exodermis, within the piliferous layer, is often impenetrable, and so the sap, after passing freely through the endodermis, may be held up at the exodermis, with the result that cork-formation occurs in the outer cortical layers. In illustration of their view, Priestley and Woffenden[1] quote the case of the Aroid, *Monstera deliciosa*, Liebm. In an aerial root twenty feet long, a "primary" endodermis was found within a few inches of the tip, while 27 inches from the tip, phellogen, which had formed two or three layers of cells, was visible beneath the exodermis. At the base of the root, six to eight layers of cork cells were present, and a thick band of sclerenchyma had also formed outside the endodermis, but the endodermis itself was still "primary" and permeable.

Priestley's interesting and suggestive theory seems to fit in with many of the known facts, but it remains for future research to determine whether it can be applied universally.

It seems to me conceivable that, if Priestley's theory of the permeable endodermis be finally adopted to explain the lack of pericyclic phellogen in Monocotyledonous roots, it may, in addition, be found to explain both their frequent possession of a storage cortex, and also their lack of secondary thickening, since the root-cambium of Dicotyledons is, at its first initiation, a pericyclic meristem. It is perhaps significant that both in the roots of *Dracaena*[2] and in the root-bases of *Dioscorea*[3]—the only two examples at present known of secondary thickening in Monocotyledonous roots—the cambium is purely cortical, although the formation of the daughter-roots in *Dracaena* is due to a pericyclic meristem[2], whose development has the effect of disrupting the endodermis[4].

A curious feature, which occasionally occurs in Monocotyledonous roots, is a splitting or dichotomy of the stele, followed by a corresponding bifurcation of the root as a whole—a process which may even be repeated more than once. This has been recorded for the Date-palm, *Phoenix dactylifera*, L.[5]; it is, however, a totally different phenomenon from the basic polystely, passing into monostely towards the root-tip, which has been described for a number of Palms[6]. Fig. xii illustrates the only case I have

---

[1] Priestley, J. H. and Woffenden, L. M. (1922).   [2] Lindinger, L. (1906).   [3] *Ibid.* (1907).
[4] Mann, A. G. (1921).   [5] Buscalioni, L. and Lopriore, G. (1910).   [6] Drabble, E. (1904).

myself met with—that of a schizostelic radicle in one of the Liliaceae, *Ruscus aculeatus*, L. The seedling itself is shown in Fig. xii, 1; the primary root, *r.*, is cylindrical near the base, but becomes deeply grooved lower down, and finally splits into two components, *r'.* and *r'.* Figs. 2 A—C show the anatomical changes which accompany the derivation of the two roots from one. The stele near the base of the radicle is 10-arch; there is, however, nothing abnormal in this, for in two other seedlings which I had the opportunity of examining, there was no forking of the radicle, and yet the steles were respectively 10-arch and 8-arch. Fig. 2 B shows the stele preparing to divide, while in Fig. 2 C the division is complete and the single 10-arch stele, which supplied the base of the radicle, is represented by two

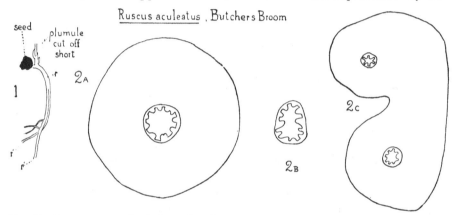

Fig. xii. *Ruscus aculeatus*, L., Schizostelic radicle. Fig. 1, seedling whose radicle, *r.*, bifurcates into *r'.* and *r'.* (× ⅓). Figs. 2 A—C, transverse sections of radicle to show bifurcation of stele; xylem and sclerised conjunctive tissue outlined within the circle of the endodermis (× 23). Fig. 2 A, close to base of radicle, root-hairs omitted. Fig. 2 B, stele alone, farther from the base of the radicle than fig. 2 A, showing process of bifurcation. Fig. 2 C, region, still farther from base, in which stele has bifurcated, but cortex is not yet fully divided. [A.A.]

steles, one of which is 8-arch while the other is 6-arch. The bifurcation of the stele becomes complete at a level at which the root shows scarcely any external sign of splitting.

Cases of steler bifurcation, such as that just described for *Ruscus*, are of interest in relation to the curious tuberous roots of the Ophrydeae, which have been the subject of a good deal of controversy. If a plant of the Early Purple Orchis, *Orchis mascula*, L., is dug up when it is in flower, its underground system will be found to consist of two tubers and of a number of slender roots (fig. xiii, 1A). The flowering axis forms the upward continuation of one of these tubers (t_1), which is slightly wrinkled and a little darker in colour than the second tuber (t_2). This distinction was noted by Sir Thomas Browne in the seventeenth century, and he adds, "the one which is fullest shootes." Fig. xiii, 1B, which represents the base of a plant cut longitudinally, justifies this early observation, for it shows that the bud

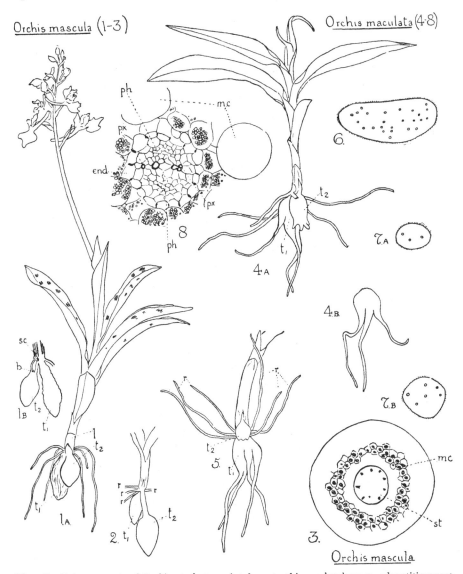

Orchis mascula (1-3)

Orchis maculata (4-8)

Orchis mascula

Fig. xiii. Tuberous roots of *Orchis*; t_1, last year's tuber; t_2, this year's tuber; $r.$, adventitious root. Figs. 1—3, *Orchis mascula*, L. Fig. 1 A, habit drawing of flowering plant ($\times \frac{1}{2}$), Chiddingly, Sussex, April 12; the scale-leaf $l.$ is burst open at the base by the dropper t_2. Fig. 1 B, base of same plant cut longitudinally; $b.$, apical bud of tuber for next year ($\times \frac{1}{2}$); $sc.$, reduced, closed scale-leaf, the base of which is continued into the dropper. Fig. 2, base of non-flowering plant, Chiddingly, Sussex, April ($\times \frac{1}{2}$). Fig. 3, transverse section of slender adventitious root ($\times 23$); $st.$, stele with 8 protoxylem groups; $m.c.$, middle cortex whose cells contain clumped fungal hyphae. Figs. 4—8, *Orchis maculata*, L. Fig. 4 A, habit drawing of a young plant which would have probably flowered later in the season, Chiddingly, Sussex, April 16; spots on leaves omitted ($\times \frac{1}{2}$). Fig. 4 B, t_1 in fig. 4 A, viewed from the other side ($\times \frac{1}{2}$). Fig. 5, base of a plant similar to that drawn in fig. 4 A ($\times \frac{1}{2}$). Fig. 6, transverse section of a tuber of the current year, showing numerous steles ($\times 5\cdot6$). Figs. 7 A and B, transverse sections of two tails of tubers of last year with respectively 3 and 6 steles ($\times 5\cdot6$). Fig. 8, transverse section of one of the steles from a tail of a last year's tuber ($\times 77$); $px.$, protoxylem; $ph.$, phloem; $end.$, endodermis; $m.c.$, mucilage cell. [A.A.]

for next year (*b*.) is located at the top of the unwrinkled tuber, t_2. The mode of connexion between the new tuber and the axis from which it is derived is a curious one. A bud arises in the axil of one of the lower leaves of the flowering shoot; the first leaf, or prophyll, of this bud is a reduced structure, which remains attached to the parent axis; it has an extremely inconspicuous free apical point, and the only part which is at all noticeable is the base, which is prolonged downwards into a tube, like a glove-finger, which bursts through the base of the axillant leaf and carries the root-tuber and the apical bud down into the earth. This arrangement appears to be comparable with that of the "droppers" of Tulip bulbs, which we shall consider later (p. 150). Fig. xiii, 2, p. 27, represents the most dropper-like tuber-stalk (prophyll base) which I have seen in *Orchis mascula*; it belonged to a non-flowering, two-leaved plant, and the tuber-stalk was 14 to 15 mm. long.

Orchis maculata, L., the Spotted Orchis, has again a pair of tubers, but they differ from those of *O. mascula* in terminating below in tail-like projections, which resemble, in their thin apical regions, the normal adventitious roots. Anatomical examination of the tubers reveals nothing but root-structure, except in the apical leaf-bud region. Transverse sections show that, both in *O. mascula* and *O. maculata*, the tubers are polystelic, being traversed by a number of vascular cords, each of which is equivalent to the single central cylinder of a slender adventitious root (figs. xiii, 6, 7, 8): the steles vary a good deal in size, and may be diarch, triarch or hexarch. It is probably the external appearance of the tailed tuber of *O. maculata* which has led to the generally accepted view that the tubers arise through the concrescence of adventitious roots. At first glance, this theory may seem to have a good deal to commend it, but it scarcely stands critical examination. If the tuber of *O. maculata*, for instance, arose through root fusion, we should expect that, when it breaks up distally into slender elongated tails, each of these tails would represent one of the roots which had originally fused. But we find, on examination, that the tails have not the structure of a normal single root—a four-tailed tuber, for example, was found to possess in its upper region sixteen steles; and the two tails shown in figs. xiii, 7A and B, contained respectively three and six steles. Such an argument as this is not, however, conclusive; and it is, indeed, hardly possible to clinch the matter from a consideration of *Orchis mascula* and *O. maculata* alone. The final elucidation of the question has come from a Canadian writer, J. H. White[1], who has been able to draw on a wider field of comparison. He has shown that, in certain Ophrydeae, the slender lateral roots—as well as the tubers—are polystelic. The tip of each of these polystelic, non-tuberous roots is enclosed in a single root-cap, and

[1] White, J. H. (1907).

there is no indication of ontogenetic fusion. In these lateral roots, the steles may either be derived by segmentation of the original monostele, or the division may be carried farther back, so that the steles arise independently from the vascular tissue of the stem. In the tubers, on the other hand, most of the steles are free from the beginning, but some arise by bifurcation of existing steles. In the case of a species of *Habenaria* (*H. hyperborea*, R.Br.), White found that the tubers of a very early generation were monostelic at the extreme base of the root, while, nearer the apex, the stele broke up. It seems impossible to reconcile these facts with the concrescence theory, but they fit quite well with the view that the root-tubers of the Ophrydeae are single roots of polystelic structure.

A feature of the Orchidaceae, which seems to bear intimately on the development of these tuberous roots, is the association of the slender adventitious roots with fungal hyphae. Fig. xiii, 3, p. 27, shows a transverse section of one of the lateral roots of *Orchis mascula*, whose stele (*st.*) has eight protoxylem groups: the cells of the middle cortex (*m.c.*) contain clumped fungal hyphae. This association of fungus and root—for which "mycorrhiza" is an inexact but convenient term—is practically universal in the Orchidaceae[1]. We owe to the French botanist, Noël Bernard, a study of the relation of the Orchids and their infesting fungi, which is a model of delicate observation, close reasoning and lucid exposition[2]. Bernard points out that, as fungal infection is the rule for the Orchid family, it is no use to try to determine what are the symptoms of infection by comparing plants of the same or related species. But, among the Ophrydeae, the individual plant in the course of its life frees itself from infection at well-determined epochs, and it is thus possible to compare the sequence of its development before and after infection. Bernard finds that, in the Ophrydeae, infection regularly results in "tuberisation," which is a mode of development characterised by "a retardation in the histological and morphological differentiation of the growing points or buds, coinciding with a laying up, in the form of reserve, of the food not utilised for differentiation." Bernard chose *Orchis montana*, Schmidt (*Habenaria nigra*, R.Br.), as a type in which to follow the life-history of the Ophrydeae. In the late summer, at a period when the plant is free from fungal infection, differentiation takes place: the leaves and the inflorescence which will unfold next year are laid down, and buds are developed in the axils of the lower leaves. At first these buds are perfectly normal, but tuberisation of the one which is youngest and least developed begins as soon as the first absorbing roots enter the earth and become infected by soil-fungi; the other buds commonly die, but they may live and become tuberous also. From the moment at which the infection of

---

[1] See Ramsbottom, J. (1922), for a critical recent account of the subject.
[2] Bernard, N. (1902), (1904), (1909); see also Magrou, J. (1921).

the root makes its influence felt upon the bud which is destined to survive, nearly all the nutriment flowing to it is laid down in a reserve form, instead of being used up in the process of differentiation; indeed, such buds at the end of May are scarcely more advanced than in the preceding September. Bernard gives evidence which indicates that, when one of the Ophrydeae is *not* infected—an extremely rare occurrence—there is no tuberisation, and the bud, which would have formed a tubercle, elongates into a branch; the fact that these buds are capable of developing into normal branches, authorises us in accepting the view that the formation of tubercles is teratological. The life-history of the plant is thus sharply marked off into two epochs: (i) a short period of normal differentiating growth in the late summer, coinciding with the absence of fungal infection, and (ii) a long period of tuberisation, beginning in the autumn, and lasting until the following summer, which coincides with the period of fungal infection of the roots. In other words, the life-history of the Ophrydeae is subject to a rhythm, in which periods of infection and non-infection alternate. The terminal bud isolated with each tubercle is free from infection, and has time to differentiate and to form its future inflorescence and its lateral bud (next year's tubercle) before the absorbing organs develop and become infected, and begin to exert their retarding, tuberising influence on the bud. As soon as the disappearance of the roots frees the bud from the "action at a distance" exerted by the fungus, it resumes its normal and rapid development. We will consider in a later Chapter (p. 221) the question of the actual means by which the influence of the fungus is exercised.

The feature in the life-history of the Ophrydeae which renders possible the alternation of infected and non-infected periods, is that every year each plant in great part disappears, only a single bud with a tubercle surviving to the next season. An example of a contrasting kind is however furnished by the Bird's-nest Orchid, *Neottia Nidus-avis*, Rich., in which infection is a permanent and continuous feature. The life-history has been elucidated in detail by Bernard, from whose work the following account is derived. The adult plant is found from May to July, growing under the shade of high beeches, where the soil is permanently covered with dead leaves. The erect floral axis of *Neottia* only bears some leaves reduced to sheaths, and flowers in a spike. If this flowering axis is followed downwards, it is found to be united underground to a rhizome entirely hidden by thick fleshy roots, crowded against one another (fig. xiv, B), the whole forming a compact, bird's-nest-like mass, to which the plant owes its name. *Neottia* is non-chlorophyllous and incapable of assimilating; it draws all its nourishment from the humus in which it lives, and develops almost entirely in darkness, pursuing its career underground for from seven to eleven years before raising its flowers into the light. The fungus enters the seedling at a very early stage, penetrating from the end where the embryo was attached to the suspensor, so that even minute plantlets, 0·25 to 0·33 mm. long, are found to be infected. Tuberisation begins when the seedling is little more than 2 mm. long (fig. xiv, A); a swelling like a pin's head arises in the apical region of the embryonal axis, and from this the first roots appear as exogenous outgrowths.

Neottia furnishes an example of an unusual mode of reproduction—budding from detached roots. Roots, which by the death of a rhizome, have become isolated in the soil, are fungus-infected in their posterior region, the growing apex alone being free. This meristematic tip, by active cell-

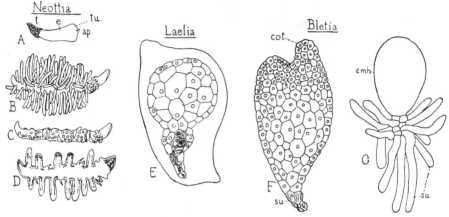

Fig. xiv. A—D, *Neottia Nidus-avis*, Rich. [Bernard, N. (1902)]. A, seedling (× 4) in May, which germinated previous June or July; *t.*, testa; *e.*, embryonal axis; *tu.*, region of tuberisation; *ap.*, terminal bud. B—D, rhizome dug up in May, apex to right; B, external appearance; C, appearance after removal of roots, showing scale-leaves; D, diagrammatic longitudinal section showing apical and lateral buds, scale-leaves, and the extension of the infested zone, which is dotted. For clearness the bases of the roots are represented on a disproportionately large scale. E, *Laelia* (garden hybrid) [Bernard, N. (1902)]. Optical section of embryo still enclosed in testa (× 50); the swollen region at the pole away from the suspensor is green. The fungus only enters from the suspensor pole. F and G, *Bletia hyacinthina*, R.Br. [Bernard, N. (1909)]. F, longitudinal section of embryo from ripe seed; fungus absent; *su.*, withered suspensor; *cot.*, cotyledon. G, young embryo and branched suspensor, two months before maturity of seed, on same scale as F.

division, produces a little milk-white tubercle, which bursts open the root-cap, and finally develops a terminal bud. Reproduction by root budding is also known in other Orchid genera, e.g. *Habenaria*[1].

The endophyte which infects *Neottia*—whether it attacks a seedling or an individual developed from a root-bud—propagates itself ceaselessly from cell to cell, inhabiting the middle cortex of rhizome and roots (fig. xiv, D). The infected region is in continuity throughout the whole body of the plant, and, normally, has no region of contact with the external surface.

The association of members of the Orchidaceae with fungi throws light on a certain curious characteristic of Orchid seedlings—the lack of a primary

[1] Lindinger, L. (1908).

root. The radicle of Flowering Plants is, as a rule, differentiated at the suspensor pole of the embryo. But in Orchid seedlings the infection takes place through this pole—a feature which is shown for *Laelia* in fig. xiv, E —and since the cells penetrated by the endophyte lose the power of differentiation, no primary root can be formed.

It is not unusual for the slender adventitious roots of Ophrydeae to be negatively geotropic (*r.*, fig. xiii, 5, p. 27). This is a character which is still

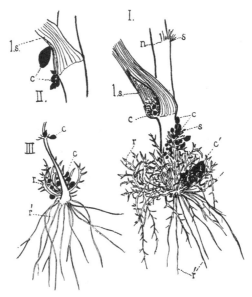

Fig. xv. *Moraea ramosa*, Ker-Gawl. Herbarium specimens from S. Africa (Brit. Mus. Nat. Hist.) to show spiny roots and cormlets (× ½). I and II from the same plant; I, base of axis, showing basket-work of spinous roots, *r.*, and descending absorptive roots, *r'.*; *c.*, cormlets; *c'.*, large lateral corm; *s.*, stalks from which the cormlets have fallen; *n.*, node; *l.s.*, leaf-sheath split open by the pressure of a cluster of corms. II, the node above the highest node in I, to show a group of cormlets of various ages, *c.*, which have burst through the median region of the base of the leaf-sheath, *l.s.* III, base of a stem, with half the basket-work cut away; lettering as in I and II. [A.A.]

more marked in certain other Monocotyledons. *Moraea ramosa*, Ker-Gawl., is a striking instance. This plant possesses both slender positively geotropic roots (*r'.*, figs. xv, I and III) and also stiff roots, *r.*, whose laterals are converted into spines; most of these spinous roots grow upwards, forming a prickly basket-work round the base of the stem. Amidst the tangle of spiny roots, there are stem-branches bearing small corms, and it has been suggested by Scott[1], to whom we owe a study of a species of *Moraea*, that the spinous roots perhaps serve more especially for the protection of the corms. But the examination of herbarium material of *M. ramosa* shows that, at

[1] Scott, D. H. (1897).

least in this species, an axillary group of cormlets is characteristic of a whole succession of nodes, and it thus seems that the association of one or more of these groups with the basal basket-work can scarcely be other than fortuitous.

It is conceivable that, in such plants as *Moraea*, there may be some connexion between the unusual spinous character of the roots and their negative geotropism, since both these features are again found associated in the roots of a Palm, *Acanthorhiza aculeata*, H. Wendl.[1] These characters will probably prove, in the ultimate analysis, to be symptoms—on the one hand, structural, and on the other, physiological—of some peculiarity in the chemistry of the organ, for recent work seems to indicate that differences in geotropic behaviour are related to chemical differentiation.

[1] Velenovský, J. (1907).

CHAPTER III

THE AXIS

IN the study of the root, which occupied the preceding Chapter, we were dealing with an organ which is morphologically simple and homogeneous, and which normally gives rise only to organs like itself. The position is however markedly different when we turn to the shoot. In the general acceptation of the term, the shoot consists of an axis bearing organs belonging to another morphological category—namely, leaves. According to this simple conception, the internodes are purely axial, and the two morphological units—stem and leaf—come into connexion at the nodes only. But there are obvious flaws in this notion, and, at different times, other ideas about the shoot have held sway; these ideas have recently culminated in an hypothesis—the Leaf-skin Theory of E. R. Saunders[1]—which seems to transform the academic conception of the leaf-bearing axis into something which approximates much more closely to reality. On this theory, the entire stem-surface consists of areas, each of which belongs to an individual leaf and can be traced upwards to the exsertion of that leaf. We will consider the subject more fully when the leaf itself comes under discussion (p. 91); the point which concerns us now, is that the leaf-skin theory precludes a belief in the existence of a naked stem, and the problem then arises: if the surface of the shoot is foliar, and the core of the shoot is axial, where does the limit come between leaf and axis? This difficult question must, I think, be left open for the present, and, for the answer to it, we must look to future research. I have headed this Chapter, "The Axis," in deference to the accepted units of morphology, but since there is actual continuity of tissue between the whole surface of the axis and the enveloping foliar mantle, it becomes difficult and artificial to treat the axis without reference to the leaf. As a matter of convenience, I shall try, however, in the following pages to focus attention as far as possible on the axial part of the shoot, and to leave the foliar part for future consideration (pp. 54–155). But I feel that the idea of *the shoot* is a much more fruitful conception than that of *the axis*, and I am not without hopes that, in the morphology of the future, it may become possible to treat the shoot as a unit, with the axis and leaves as subsidiary elements—just as we now treat the leaf as a unit, while recognising the distinctness of the leaf-base, petiole and lamina.

Owing to the lack of secondary thickening, the primary vascular system of Monocotyledonous stems achieves a greater degree of permanence and

---

[1] Saunders, E. R. (1922).

importance than does that of Dicotyledons; from its initiation, also, it
occupies a greater proportion of the axis. In Dicotyledons, the procambial
layer, in which the leaf-bases are differentiated, generally takes the form of
an elongated hollow cylinder, in which the traces follow a more or less
vertical course, but in Monocotyledons, the procambium often lies comparatively near the surface and—in correlation with the slowness of growth
and the crowding of the nodes—it very commonly tapers rapidly upwards,
so that its form becomes markedly conical rather than cylindrical. The
conical form of the procambial mantle leads directly to that peculiar
curvature of the leaf-traces which is a noticeable feature of so many

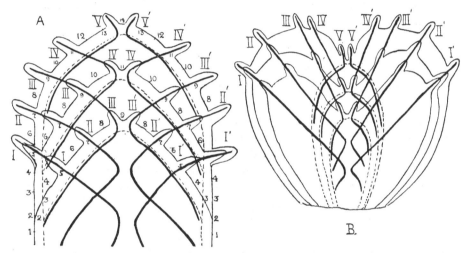

Fig. xvi. Diagrams to illustrate course of leaf-traces in Monocotyledonous axes; founded on description
in Mohl, H. von (1858), and Schoute, J. C. (1903). In both A and B the history of leaves I—V and
I′—V′ are traced through three stages; procambial mantle, dotted; leaf-traces, heavy black line.
A, stem with conical growing apex; points marked with numbers 1, 2, 3, 4 etc. correspond in the
three stages, and show where growth takes place. B, stem with abbreviated axis and depressed
growing point. For interpretation of diagrams see text. [A.A.]

Monocotyledonous stems. An explanation of this curvature was given by
von Mohl[1] as long ago as 1858, but his account seems to have been little
studied, probably because—being unillustrated—it is somewhat laborious
to follow. In more recent years, Schoute[2] has cited von Mohl's description,
and given illustrations of his own, in order to elucidate it, but his diagram
of the developmental stages of a vascular bundle (l.c., Pl. 4, fig. 4) does not
seem to me altogether convincing. I have therefore made an attempt in
fig. xvi to represent, in an extremely diagrammatic fashion, my interpretation
of the sequence of events which, according to von Mohl, brings about the
curious inward and outward curvature of the leaf-traces in a Monocotyledonous axis. In fig. xvi, A, I have shown three successive stages of development

[1] Mohl, H. von (1858).   [2] Schoute, J. C. (1903).

of the tip of a stem, whose growing apex is massive and conical. The procambial mantle, which lies not far from the surface, is represented by a dotted
line, while the leaf-traces I—v and I'—v' are shown as heavy black lines. The
leaves are indicated as mere protuberances, and no attempt is made to give
them their proportionate size. The youngest stage in the leaf and its trace
may be seen in leaves v and v' at the top of the diagram; the traces which
supply them are laid down on the inner side of the hollow procambial cone,
and conform to its curvature. The later history of such traces can best be
understood by following their position in the apical cone at its three successive stages. As examples we may take the leaves III and III', which
are developed at the tip of the growing point at its first and lowest stage;
these traces, like those of v and v', have, in their first phase, a simple
course, following the curvature of the cone. I have numbered off the outline of the cone at equidistant levels (1—9), and it will be seen that, at the
second stage, the apical growth has carried the point 9 away from the
extreme tip, which is now occupied by point 11. The result of this elongation, in association with the outward movement of the procambial mantle,
has been that the traces of leaves III and III', which, in the former stage,
approached one another near the apex of the stem, now diverge from this
point; the characteristic inward and outward curvature is thus established.
It will be seen from the diagram that, when a series of leaves develops in
this way, the result is a geometric pattern of interlacing curves.

Fig. xvi, B, shows a second variant on the same type of leaf-trace
system—that in which the growing point is depressed and crateriform. But
even in such stems the procambial mantle is not itself dissimilar in shape
from that shown in fig. xvi, A; the difference lies in the fact that, instead of
producing, externally, a minimal and uniform amount of parenchyma, as
in fig. xvi, A, it gives rise to a layer, which is insignificant at the extreme
apex, but immediately suffers a rapid increase, so that the actual tip is, as
it were, sunk in the centre of a parenchymatous ring-cushion. In some
cases, still farther from the growing point, the parenchyma again diminishes,
giving the turnip-shaped stem shown in fig. xvi, B (cf. *Brodiaea congesta*, Sm.,
fig. xxiii, 4 A, p. 46); in other cases the parenchyma-making activity is uniform except near the tip, and the form of the stem is thus more or less
cylindrical with a depressed apex (cf. *Hypoxis setosa*, Baker, figs. viii, A and B,
p. 20). A transverse section of one of these crateriform apices—that of
Cordyline australis, Hook.—is shown in fig. xxvii, 4, p. 50.

The contrast between figs. xvi, A and B, lies in the fact that, in A, the
central cylinder—and consequently the curved region of the leaf-trace—
occupies almost the whole axis, while, in B, the relative smallness of the
vascular cylinder results in the curvature being confined to a comparatively
limited basal part of the leaf-trace.

Attempts have been made to read an adaptational meaning into the outward curvature of the basal region of such leaf-traces as those shown in fig. xvi; Haberlandt[1], for instance, says that "in accordance with the rules of inflexible construction...these mechanical strands tend to approach the periphery." But any such teleological view is ruled out of court, when we realise that this outward curvature is a purely automatic result of the massive conical form of the procambial layer.

It must be clearly understood that the diagrams in fig. xvi are highly simplified, and that many complications are ignored—such, for instance, as the fact that numerous bundles of different ages may enter the broad leaf-bases. And, even with these reservations, the diagrams in question only claim to represent one general type of Monocotyledonous axis—that type which differs most widely from the Dicotyledon; many varieties of leaf-trace system are found, some of which approximate more or less exactly to various Dicotyledonous schemes[2]. The task of following the bundles in the Monocotyledonous axis—owing to the crowding of the leaf-traces, and their varying curvature, and the frequent choking of the tissues with reserve material—is an exacting and difficult task, which botanists seem to have shirked accordingly; indeed, most of our information on the subject is many years old. It is much to be desired that some present-day worker would apply the skill and patience necessary for obtaining fresh data with the aid of modern technique.

One of the results of the system of leaf-trace arrangement which we have been considering is that the bundles of the Monocotyledonous axis are commonly scattered over the cross-section, instead of being grouped in a ring as in Dicotyledons. There are, however, many exceptions to the scattered arrangement, and a ring is frequently formed in plants in which the internodes are of some length, and the procambial mantle approximates in form to that of a Dicotyledon; examples here figured are the peduncles of *Hypoxis setosa*, Baker (fig. viii, E, p. 20), and of *Bowiea volubilis*, Harv. (fig. xxix, E, p. 52), and the young axis of *Tamus communis*, L. (frontispiece).

Fig. xvii, p. 38, illustrates the scattered bundle-system in the stem of two plants belonging to the tribe Rusceae (Liliaceae)—*Danae racemosa*, (L.) Mönch., Victor's-laurel, and *Ruscus aculeatus*, L., Butcher's-broom. The individual bundles (figs. xvii, 1 B and 2 B) have a triangular, or more or less V-shaped, xylem—a form to which we get an approximation in certain Dicotyledons, such as *Thalictrum* (fig. xviii, p. 39).

The anatomy of *Ruscus aculeatus* becomes of particular interest, since it was one of the plants used by J. Bretland Farmer[3] in his experiments on the "specific conductivity" of various woods. He found that the Butcher's-

---

[1] Haberlandt, G. (1914), p. 383.   [2] Falkenberg, P. (1876) and Bary, A. de (1884).
[3] Farmer, J. B. (1918).

broom conducts water at a rate which is, relatively, of extreme slowness. He points out that this fact throws light on the problem of why it is that this plant, whose appearance is strikingly xerophytic, yet flourishes best in damp ground, where its associates are mesophytes, or even hygrophytes. It

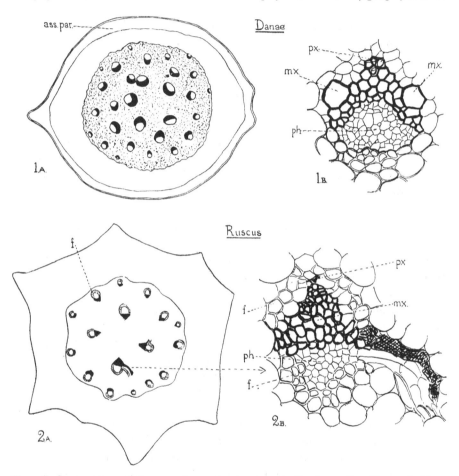

Fig. xvii. Stem anatomy of Rusceae; *px.*, protoxylem; *mx.*, metaxylem; *ph.*, phloem; *f.*, fibres. Figs. 1 A and B, *Danae racemosa*, (L.) Mönch. Fig. 1 A, transverse section of axis (×47); *ass. par.*, assimilating parenchyma; the two ridges correspond to the orthostichies. Fig. 1 B, single bundle from transverse section (×318). Figs. 2 A and B, *Ruscus aculeatus*, L. Fig. 2 A, transverse section of plumular axis (×47); the ridges correspond to the orthostichies. Fig. 2 B, bundle marked with an arrow in fig. 2 A (×318); this bundle is in the act of branching. [A.A.]

is greatly to be wished that experiments on Farmer's lines could be more widely extended among Monocotyledons, particularly among the arborescent types.

There is considerable variation in the form of the individual bundle in the Monocotyledonous axis. That of *Tamus communis*, L., the Black

Bryony, is remarkable since each bundle has two distinct phloem strands lying on the same radius of the stem. A further peculiarity of the Dioscoreaceae, which can easily be observed in *Tamus*, is that the large metaxylem vessels develop from multinucleate elements[1] (frontispiece).

A form of bundle which is frequent in Monocotyledonous rhizomes is the concentric or amphivasal strand, in which the xylem surrounds the phloem[2]. Bundles of this type are found, for instance, in two well-known Aroids, *Calla palustris*, L., the Bog-arum, and *Acorus Calamus*, L., the Sweet Flag (fig. xix, p. 40). From the accompanying figures it will be seen that the leaf-traces of *Acorus* are collateral in the cortex, though concentric in the central cylinder (cp. figs. xix, 1 A and B); the same difference is observable in the sections of *Dracaena* drawn in figs. xxi, D and E, p. 42. Jeffrey[3], who, some years ago, discussed the significance of the amphivasal

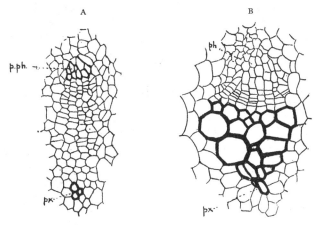

Fig. xviii. *Thalictrum flavum*, L. A, transverse section of young bundle from an inflorescence-axis gathered June 1, in which only protoxylem (*px.*) and protophloem (*p.ph.*) are fully differentiated. B, transverse section of mature bundle from infructescence-axis gathered August 23. (×320 *circa*.) [Arber, A. (1918[1]).]

strands of Monocotyledons, appears to interpret them as the result of the crowding and fusion of foliar strands at the nodes. This may be true in some cases, but there is, on the other hand, little doubt that concentric bundles are often single strands in which the xylem has enclosed the phloem[4]. The extreme type of V-bundle, in which the arms of the xylem-V embrace the phloem group, shows how easily the amphivasal condition might arise without the fusion of separate strands.

In the stems of Dicotyledons a layer of procambial tissue between the xylem and phloem of each bundle remains capable of division, and by the extension of this meristematic activity from bundle to bundle, a complete

[1] Pirotta, R. and Buscalioni, L. (1898).   [2] Many cases are cited by Möbius, M. (1887).
[3] Jeffrey, E. C. (1917).   [4] See pp. 77, 80, 81, and fig. cviii, 7, p. 139.

hollow cylinder of cambium is formed, which, by producing xylem internally
and phloem externally, is responsible for secondary thickening. In Monocotyledons, on the other hand, though the cambial zone within each bundle
is often recognisable in early stages[1], it has a short life, and its activity is,

Acorus Calamus (1)   Calla palustris (2)

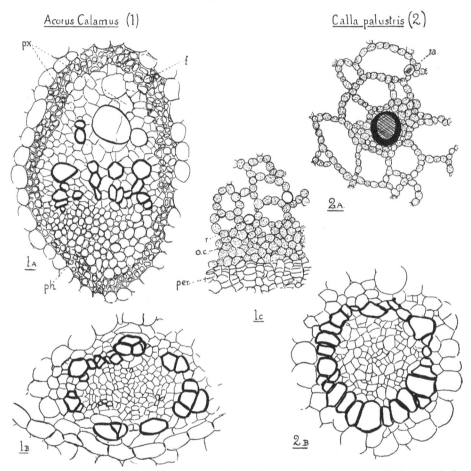

Fig. xix. Concentric bundles. Figs. 1 A—C, *Acorus Calamus*, L. Fig. 1 A, bundle from cortical
region of transverse section of stem (× 193); *px.*, protoxylem; *ph.*, phloem; *f.*, fibres. Fig. 1 B,
concentric bundle from close to stelar surface (× 193). Fig. 1 C, small segment of rhizome margin
(× 23); *per.*, periderm; *o.c.*, outer cortex; *r.*, resin sacs. Figs. 2 A and B, *Calla palustris*, L.
Fig. 2 A, small bit of rhizome in transverse section with bundle; xylem, black; phloem, cross-
hatched; *ra.*, raphides (× 47). Fig. 2 B, bundle from fig. 2 A (× 193). [A.A.]

as a rule, minimal; at maturity all trace of it is often lost (cf. figs. xx, 1 A, B, C).
But whereas intrafascicular cambium certainly exists, though it is ephemeral,
I know of no example among Monocotyledons, in which a cambial cylinder,
analogous to that of Dicotyledons, is formed by the extension of this meri-

[1] Arber, A. (1917), (1918[1]), (1919[4]).

stematic activity from bundle to bundle across the intervening parenchyma; in other words, although *intra*fascicular cambium is present, *inter*fascicular cambium is absent.

A form of cambial thickening which, though differing widely from that of

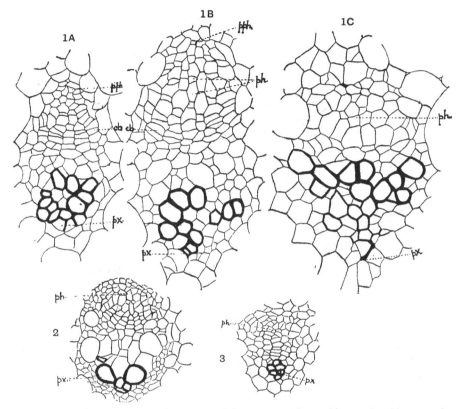

Fig. xx. Intrafascicular cambium in Monocotyledonous axes (*cb.*, cambium; *ph.*, phloem; *pph.*, protophloem; *px.*, protoxylem). Figs. 1 A—C, transverse sections of vascular bundles from inflorescence-axes of *Eremurus himalaicus*, Baker (× 275). Fig. 1 A, a bundle from the flowering region of an axis gathered May 7, showing cambium; the phloem is not fully differentiated. Fig. 1 B, a bundle from the region below the lowest flowers of the same axis, showing secondary phloem in process of differentiation into sieve-tubes and companion-cells. Fig. 1 C, a mature bundle, from the flowering region of an axis gathered on May 28, in which cambial activity has practically ceased. [Arber, A. (1917).] Fig. 2, *Acorus Calamus*, L., transverse section of single bundle from near apex of rhizome, showing radial rows of elements between xylem and phloem; only the protoxylem is at present lignified (× 320). Fig. 3, *Tamus communis*, L., transverse section of single bundle close to apex of aerial climbing stem, showing radial rows of elements between protoxylem and phloem (× 320). [Arber, A. (1918¹).]

Dicotyledons, is yet competent to produce large arboreal forms, is found, however, in certain Liliiflorae[1]; it occurs, for instance, in *Aloe, Cordyline, Dasylirion, Dracaena, Kniphofia, Nolina, Yucca, Xanthorrhoea* (Liliaceae); *Agave* and *Furcraea* (Amaryllidaceae); *Aristea*[2] and related genera[3] (Iridaceae); and

¹ Lindinger, L. (1908).   ² Scott, D. H. and Brebner, G. (1893).   ³ Adamson, R. S. (1924).

Testudinaria[1], *Tamus* and *Dioscorea*[2](Dioscoreaceae). In the slender stems of species of *Dracaena* cultivated in this country as pot-plants, one may readily observe the early phases of the thickening process. The transverse section of the aerial axis shows a large central cylinder with numerous scattered bundles, while immediately outside the fibrous outer limit of the cylinder there is a meristematic zone (*mer.*, figs. xxi, A and B), in which additional

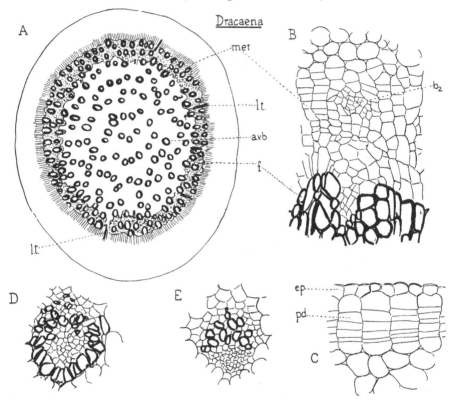

Fig. xxi. Secondary thickening in *Dracaena* sp. ("Duchess of York"). A, transverse section of erect aerial axis (× 14); *mer.*, meristematic zone; *f.*, fibrous outer zone of stele; *l.t.*, leaf-traces; *a.v.b.*, amphivasal vascular bundles. B, small segment of outer fibrous margin of stele from A, to show bundle, b_2, differentiating in the meristem (× 318). C, small outer segment of section such as A, to show periderm, *pd.*, arising in the cortical layer immediately below the epidermis, *ep.* (× 193). D, primary amphivasal bundle from stele, and E, leaf-trace from cortex, from section nearer the apex than D (× 193). [A.A.]

secondary bundles are differentiated. The rooting subterranean axis has the same form of thickening, but the boundary between primary and secondary tissues is less conspicuous, as the outer layers of the stele are not fibrous.

Thickening by the formation of these innumerable, discrete secondary strands, with intervening parenchyma, is fundamentally different from the production of a continuous secondary zone of xylem and phloem, with

[1] Mohl, H. von (1845).   [2] Queva, C. (1894).

which we are familiar in Dicotyledons. It shows, however, the same kind of periodicity, for growth-rings have been described in a number of tree Monocotyledons[1].

As compared with that of Dicotyledons, the secondary thickening of Monocotyledons tends to begin late, often lagging, as it were, behind the growth in length of the axis. The result is that the stems are sometimes quite massive at the base, but taper rapidly above[2]. *Nolina recurvata*, Hemsl., is an example, and an even more extreme case is afforded by those Dioscoreaceae in which there are slender axes which are purely primary, while the secondary growth forms a basal woody tuber; in the Elephant's-foot, *Testudinaria*, for instance, the tuber is said sometimes to form a mass of several hundredweight[3], and we may perhaps interpret this strange plant as illustrating the effect produced when the tendency to basal thickening gets out of control.

The secondarily thickened axes, which we have just considered, form only a minute proportion of the whole number of Monocotyledonous stems. In the great majority—even in such tree forms as the Palms—the primary bundle-system is permanently retained, and there are also other divergences from the type with which we are familiar in Dicotyledons.

One feature, which is characteristic of many Monocotyledonous axes, is a peculiarly slow growth in length. The result of this dilatoriness—in combination with the frequent presence of contractile roots, which exercise a downward pull—is that even those axes which are vertical often fail to emerge from the soil. Fig. viii, A, p. 20, shows this feature in *Hypoxis setosa*, Baker. Examples are also found in another genus of Amaryllidaceae—*Curculigo*; a specimen of *C. orchioides*, Roxb., in the British Museum Herbarium, has a vertical subterranean axis, which is 14 cms. long, though (in its present dry state) only 0·5 cm. in diameter. This, however, is a rare form of stem, since the Monocotyledonous axis is more frequently tuberous than elongated. There is probably some relation between the tardy development of the stem-apex and the accumulation of carbohydrate reserves characteristic of tubers and bulbs. That the lack of opportunities for translocation may, in some cases, be responsible for tuber formation is suggested by an observation of Engler's on an Aroid from Zanzibar, *Zamioculcas Loddigesii*, Decne.[4] When individual pinnules from the leaf of this species were shed in a hot-house, they each developed a little tuber at the basal end, from which roots and leaves were produced. The leaf, as Engler points out, is thick, and rich in reserves, and, as it remained green, it probably continued to assimilate after it fell. The products, which, when the leaflet was attached to the common petiole, would have passed down to the root-stock, were

---

[1] Lindinger, L. (1909), etc.   [2] Schoute, J. C. (1903).   [3] Velenovský, J. (1907), p. 584.
[4] Engler, A. (1881—1884); see I. 3, vol. i, 1881, pp. 189—190.

unable, in the case described, to get farther than the base of the pinnule, and the local accumulation resulted in tuberisation.

It would be interesting to know whether any such special explanation (e.g. an inadequacy in the means of translocation) can be found for the tuberous development of the stem in such grasses as that form of *Arrhenatherum avenaceum*, Beauv., known as var. *bulbosa*, Lindl. In this variety (fig. xxii) the lower internodes of the shoots are swollen so that they often split

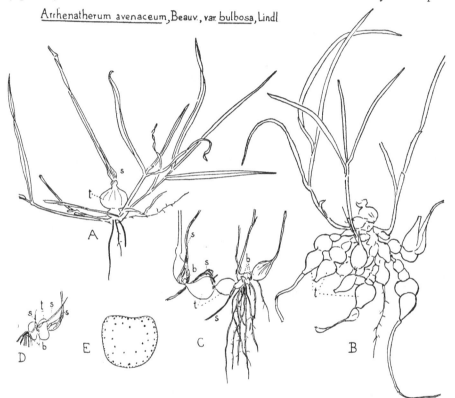

Fig. xxii. *Arrhenatherum avenaceum*, Beauv., var. *bulbosa*, Lindl. A—D, habit drawings, Chiddingly, Sussex, April, 1920 (× ½); *t.*, tuberous internodes; *b.*, lateral buds; *s.*, leaf-sheaths. E, transverse section of tuberous internode, somewhat enlarged. [A.A.]

the enclosing leaf-sheaths; they have the appearance of strings of green or brown beads. Transverse sections show that the swelling is not, as in many tubers, due to hypertrophy of the cortex, but to that of the pith, and of the ground-tissue in which the bundles lie (fig. xxii, E). An inulin-like carbohydrate, graminin, has been found in the tubers. In *Molinia caerulea*, Mönch., in which similar tuberous internodes occur, it has been recorded that the reserve material in the tubers is laid down in the form of a thickening on the cell-walls, and of protein and starch grains in the cell cavities[1].

[1] Jefferies, T. A. (1916).

Some forms of subterranean tuberous axis perennate, but in others, such as the corm, a fresh tuber is usually formed each year. In figs. xxiii, p. 46, and lxxix, 48 A, p. 105, corms are illustrated. The relations of successive corms in a species of *Cyanella* can be seen in figs. xxiii, 2 A—C. Fig. 2 A shows the corm in October, invested by a scale-leaf reduced to its fibrous veins. Figs. xxiii, 2 B and C, represent the stage reached in the following May; the corm of the previous autumn, c_1, is shrunk and shrivelled, and is surmounted by a new corm, c_2. The history of a corm can be easily followed in the Garden Crocus, in which the flowering shoots arise as buds on the old corm in the axils of scale-leaves, and, during and after anthesis, swell up at the base into the corms for next year. Figs. xxiii, 1 A—C, show a somewhat different form of corm—that of *Colchicum illyricum*, Stokes (Liliaceae), one of the species of the so-called Autumn Crocus, in which, as in the true Crocus, the new corm is lateral, but the bud for next year is carried downwards into the soil by an asymmetric growth of the old corm; as shown in Fig. xxiii, 1 A, two buds may develop in one season. I shall not attempt a full account of the life-history of *Colchicum*. For this, and for a detailed description of Monocotyledonous tubers and bulbs in general, reference must be made to the work of Thilo Irmisch (*Zur Morphologie der Monokotylischen Knollen- und Zwiebelgewächse*, 1850, and other memoirs[1]). His writings form perhaps a culmination of that exact and fully illustrated descriptive morphology, based on naked-eye observations, which flourished up to the middle of the nineteenth century, but which failed to survive the general diversion of interest due to the acceptance of the Darwinian Theory. More modern work has, to some extent, corrected and supplemented Irmisch's conclusions, but his facts, within their limits, were arrived at in too conscientious and painstaking a fashion, ever to be seriously disturbed. A recent writer, M. B. Church[2], has succeeded in overcoming the difficulties of penetration and embedding, which have hitherto prevented the application of microtome technique to the study of bulbs; it is a tribute to the accuracy of Irmisch's observations that the results obtained by these newer methods conflict, in no essential point, with his descriptions. I reproduce here certain of Church's figures illustrating the structure of the bulb which she examined—that of *Cooperia Drummondii*, Herb., the Rain-lily of Texas (figs. xxiv, p. 47, and xxv, p. 49). Whereas a corm is a thickened axis, whose associated leaves contain little reserve material, a bulb is a reduced and abbreviated axis, which itself stores little food, but is crowded with leaves or leaf-bases, whose cells are full of reserve substances. It is, as it were, a telescoped shoot, in which the internodes have almost no vertical elongation; fig. xxiv, A, shows how the bulb of *Cooperia* would appear if the reverse process took place, and the internodes were drawn out as in a normal shoot.

[1] On *Colchicum* see also Rimbach, A. (1897[4]). [2] Church, M. B. (1919).

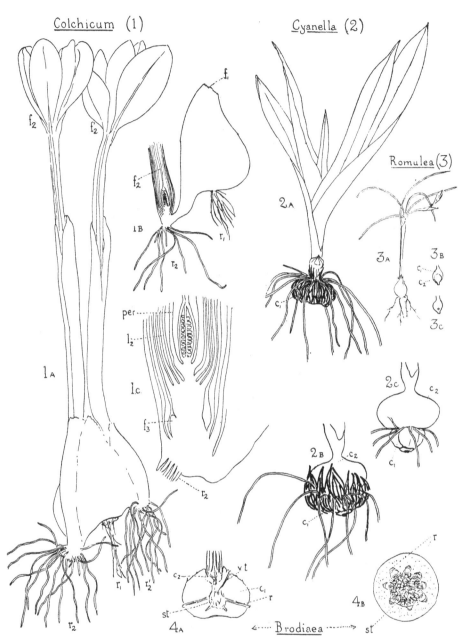

Fig. xxiii. Corms. Figs. 1 A—C, *Colchicum illyricum*, Stokes, var. *superba*. 1 A, corm (October 6, 1923) with two flowering shoots, f_2 and f'_2, and roots r_2 and r'_2 ($\times \frac{1}{2}$); r_1, roots associated with flowering shoot of 1922. 1 B, corm cut in half longitudinally, to show flowering shoot f_2; f_1, remains of flowering shoot of 1922 ($\times \frac{1}{2}$). 1 C, base of flowering shoot f'_2, cut in half and enlarged to show f_3, flowering shoot for 1924; l_2, leaves which will emerge next spring; *per.*, perianth-tube of flower f'_2. Figs. 2 A—C, *Cyanella* sp. 2 A, plant in October; c_1, corm ($\times \frac{1}{2}$). 2 B, similar plant flowering in the following May (corm alone represented) ($\times \frac{1}{2}$). 2 C, same corm as 2 B, with fibrous corm-cover removed; c_1, shrivelled corm of previous year; c_2, new corm enclosed in sheathing leaf-bases. Figs. 3 A—C, *Romulea Columnae*, Sebast. et Mauri, Dawlish Warren, April 23, 1918 ($\times \frac{1}{2}$); 3 B, new corm c_2 attached to old corm c_1; 3 C, old corm from which new corm has been removed, to show small circular area of detachment. Figs. 4 A, B, *Brodiaea congesta*, Sm., early spring. Fig. 4 A, corm cut longitudinally ($\times \frac{1}{2}$); c_1, last year's corm; c_2, this year's corm; *v.t.*, vascular tissue for last year's growing point; *st.*, stele; *r.*, root; fig. 4 B, transverse section of corm ($\times \frac{1}{2}$). [A.A.]

But in the actual bulb, the sheaths of the successive leaves arise practically at the same horizontal level, and are enclosed one within the other; this is shown as a transparency in fig. xxiv, B, and in section in fig. xxv. The excessive reduction in the internodes of bulbs is naturally associated with a minimal growth in length. Rimbach[1] has recorded, for instance, that the bulb axis of *Stenomesson aurantiacum*, Herb., grows 5 mm. annually, while the flowering shoot, on the other hand, is 40 cms. high. But, definite as are the characteristics of the extremest form of bulb, it is artificial to try to

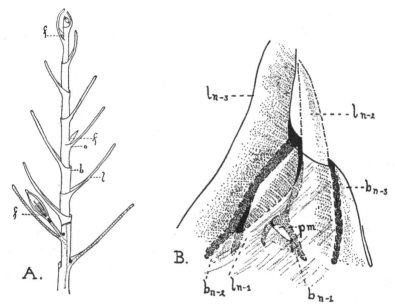

Fig. xxiv. A. Diagram of bulb of *Cooperia Drummondii*, Herb., as it would appear if the internodes developed; *l.*, limb of a leaf; *b.*, basal sheath; *o.*, leaf subtending flower; *f.*, flower; in the lower part of the figure a central longitudinal section is represented, showing principally the relation of the basal sheath to the main axis; on the evidence to hand, the relation of the flower and its subtending leaf to the main axis is problematic. B. Diagrammatic semi-transparency of the youngest portion of the bulb of *Cooperia Drummondii*: *pm.*, the primordial area, below which is the primordium of the youngest or *n*th leaf; successively older leaves are marked l_{n-1}, l_{n-2}, and l_{n-3}; their respective bases, b_{n-1}, b_{n-2} and b_{n-3}. [Church, M. B. (1919).]

establish a hard-and-fast distinction between bulbs and other subterranean forms of reserve-containing shoots. Some species of *Allium*, for instance, have rhizomes, while others have bulbs, but, as Irmisch[2] has pointed out, these rhizomes are nothing more than the bulb axes of many years' growth, which have lost their leaves. We do not know what determines whether it shall be one organ rather than another which becomes loaded with food material and hence suffers a distortion of form; the evidence of storage

[1] Rimbach, A. (1896²).   [2] Irmisch, T. (1850).

conditions in related plants suggests that the accumulation of reserves may readily be deflected from one region to another. Among the inflorescence bulbils of *Globba* (Zingiberaceae), for instance, we find certain species in which storage takes place in root-tubers, while in others there are storage rhizomes, and the roots are not tuberous[1]. Again, in the Bambuseae—as in other Grasses—the endosperm is normally the storage tissue for the seed, but in *Melocanna* the ripe seeds are endospermless, and it is the scutellum and pericarp which contain the reserve[2].

We have so far considered only those Monocotyledonous stems whose direction of growth is vertical and negatively geotropic, but a large proportion of the less abbreviated axes take a horizontal and often subterranean course—only the foliage leaves and the reproductive stems rising into the air. The accompanying drawing of *Cladium Mariscus*, R.Br., is an example of a horizontal axis bearing scale-leaves; when the growing apex turns upwards, the development of the underground axis is continued, in the established line, by a lateral bud. In certain Orchidaceous saprophytes there are no roots, and all the absorption is carried out by the subterranean shoots. In one of these cases, *Epipogum aphyllum*, Sw., for instance, the rhizome-axis bears absorbing hairs, whereas in *E. nutans*, Reichb., the surface of the rhizome is smooth, and the intake of water seems to be due to long hairs produced by the scale-leaves[3].

But the most peculiar of Monocotyledonous axes are perhaps those which are not only subterranean, but which grow vertically downwards into the soil. Stolons of this nature have long been known in the genus *Cordyline*[4]. If a pot-plant of *C. australis*, Hook., such as that shown in fig. xxvii, 3, p. 50, is uprooted, it presents, at first glance, a most puzzling appearance; it seems to have a vertical subterranean axis, with a growing point at the surface of the soil, and a second opposite growing point at the lower extremity (*st.*). But the examination of young stages puts another complexion on the matter. Fig. xxvii, 1, shows the base of a young plant in which the axis is extremely abbreviated and bears a rosette of leaves. There is no sign of any downward portion. But in fig. 2, which represents a slightly older plant, a stolon, *st.*, has arisen laterally and has burst through the base of the leaf, *l*. When we examine fig. xxvii, 3, in the light of these earlier stages, we notice that the downwardly directed stem is not absolutely in a straight line with the leaf-bearing upper portion; it is no doubt merely an overgrown lateral stolon, which, on account of its elongation, looks more important than the dwarfed parent axis.

A still more singular mode of positively geotropic growth is found in certain Palms belonging to the genus *Sabal*. Here the axis in the earlier

[1] Leeuwen, W. D. van (1921).   [2] Stapf, O. (1904).
[3] Groom, P. (1895).   [4] Lindinger, L. (1908), etc.

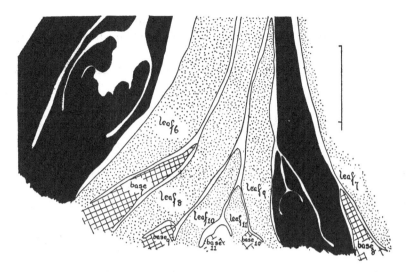

Fig. xxv. *Cooperia Drummondii*, Herb., longitudinal section of youngest portion of mature bulb; black areas represent the flower and its subtending leaf; stippled areas indicate the portions of the leaves which become the limbs; cross-hatched areas show the leaf-bases; plain areas are primordial tissue; it is presumed that the oldest leaf shown is the sixth leaf—an arbitrary choice. Scale = o·1 mm. [Church, M. B. (1919).]

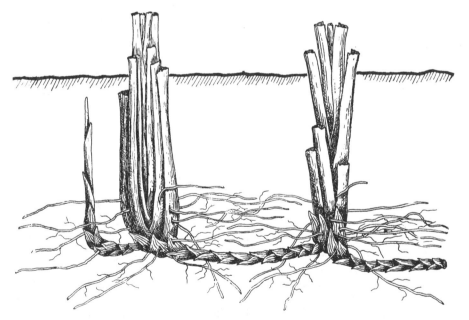

Fig. xxvi. *Cladium Mariscus*, R.Br., in moderately damp soil, Wicken Fen, September (× ⅜). [Yapp, R. H. (1908).]

stages grows downwards, and the result, in combination with the upward growth of the leaves, is to produce a crook-like form, in which the scar of the primary root may even be raised into the air[1].

As compared with Dicotyledonous axes, those of Monocotyledons are

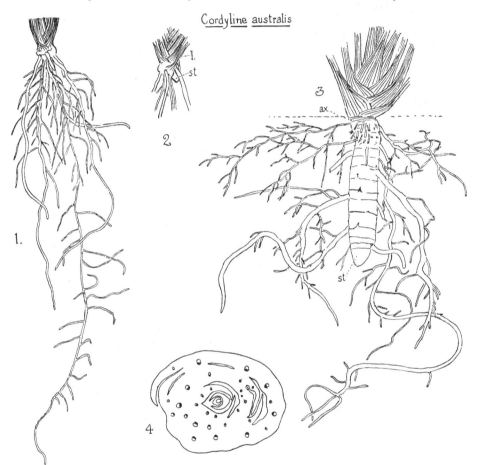

Fig. xxvii. Development of a positively geotropic stolon in *Cordyline australis*, Hook. 1, root system and base of shoot of young plant (× ½). 2, base of slightly older plant, showing downwardly directed stolon, *st.*, bursting through base of leaf, *l.* (× ½). 3, underground system and base of aerial shoot of an older plant (× ⅓). The region *ax.* of the stem, from which the aerial leaves spring, is probably the original stem, but the main part of the underground axial region consists of a large stolon, *st.* 4, transverse section of central part of stem-apex (× 14). [A.A.]

poorly branched. It is probable, however, that this does not point to any inherent incapacity for branching. It is more likely that it is merely a mechanical effect of the characteristic abbreviation of the axis, and its enclosure within a succession of leaf-sheaths; these factors result in a lack of space

---

[1] Karsten, H. (1847), p. 82; Baillon, H. (1895), pp. 251—2; Velenovský, J. (1907), p. 593.

for bud-development, and the formation of branches is thus confined and hindered. An occasional minor outcome of the breadth of the leaf-bases is the occurrence of a series of col-
lateral buds in each leaf-axil, instead of a single bud. Fig. xxviii shows the development of numerous axillary buds in the bulb of *Allium nigrum*. Figs. xxviii, I A, B, represent the bulb in its December state, while, six months later, the shrivelling and disappearance of the outer scale-leaf, sc_1, reveals a circle of bulblets, *ax.*, which have developed in its axil (fig. 2). Sometimes the conditions of pressure under which the lateral buds arise result in anomalous fusions; in *Ornithogalum caudatum*, Ait., for example, the pedicels of the numerous bulblets unite with one another and with the face of the sheath of the succeeding leaf[1].

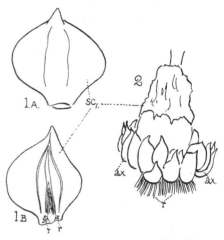

Fig. xxviii. 1 and 2, *Allium nigrum*, L. (× ½). 1 A, bulb in December; 1 B, longitudinal section of the same bulb; *r.*, adventitious roots. 2, bulb in the following June; the thick outer scale, sc_1, has shrivelled, and all its lower region has disappeared, revealing a complete ring of buds, *ax.*, which have developed in its axil. [A.A.]

Although in a large proportion of Monocotyledonous shoots the axis is abbreviated and the leaves are more conspicuous organs than the stem, there are also many members of the class in which the axis is elongated and slender, while the internodes are long. For instance, in *Daemonorops Draco*, Bl. (Palmae), the axis, which has a diameter of only 3 cms., may exceed 150 metres in length[2]. And not only may the stems be elongated, but the leaves are sometimes reduced to small scales, so that the shoot seems, at first glance, to consist of axes alone. This peculiarity is shown by *Bowiea volubilis*, Harvey, a Liliaceous climber from S. Africa; here there are some linear basal leaves (*l.*, fig. xxix, B, p. 52), but the climbing axes, which are green assimilating organs, bear only scale-leaves (*sc.*), and their ultimate branches terminate in flowers, which are often abortive.

A case which is related to that of *Bowiea*, but which I include here with some reserve, because of its controversial character, is that of the "needles" of *Asparagus*—the organs which are responsible for the feathery appear-
ance of the mature shoot, and hence for the florist's name of "Asparagus Fern." The "needles" are found in clusters, often associated with flower pedicels, in the axils of scale-leaves (fig. xxx, p. 53). They have sometimes been regarded as foliar, but a recent study of them[3] has convinced me that the more widely-accepted view, which treats them as axial, is the correct one.

[1] Lonay, H. (1902).   [2] Velenovský, J. (1907).   [3] Arber, A. (1924³).

Those who have upheld the axial theory, have, however, invariably regarded
the needles as *naked* axes. Until recently such a conception has presented
no particular difficulty, but the leaf-skin theory, which has revolutionised
our idea of the shoot, has made it highly improbable that any axes among
the higher plants are truly naked. This seemed to me, at first, to put
a difficulty in the way of accepting the axial theory of the *Asparagus*

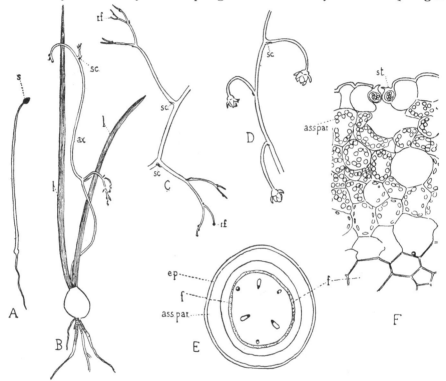

Fig. xxix. *Bowiea volubilis*, Harv. A, seedling (× ½); *s.*, seed. B, young plant with foliage leaves, *l.*,
and climbing axis, *ax.* (incomplete), with scale-leaves, *sc.* (× ½). C, sterile region of climbing shoot
(× ½); *r.f.*, rudimentary flowers. D, fertile region of climbing shoot (× ½). E, trans. sect. of peduncle
(× 23); *ep.*, epidermis; *ass.par.*, assimilating parenchyma; *f.*, fibrous zone; F, part of the outer
region of E to show assimilating tissue with stomate, *st.* (× 193). [A.A.]

needle; but when I came to examine these organs in detail, I found that
the difficulty had no real existence. For the needles cannot, strictly, be
described as "naked," since certain of them, in *Asparagus Sprengeri*, Reg.,
bear minute scale-leaves, which seem to have been hitherto overlooked[1].
The drawings from serial sections of young shoots in figs. xxx, 1 and 2,
show these structures; in figs. 1, D and E, each of the needles, W and X,
bears minute and delicate non-vascular scales, l'_1, l'_2, and l'_3. The discovery
of these leaves removes the only serious objection to the axial theory, and

[1] During the passage of this book through the press, I have been informed by Prof. L. Buscalioni
that he recorded the existence of these structures in 1914 (Boll. d. Acc. Gioenia d. Sci. Nat. in Catania,
Fasc. 31, Ser. 2 *a*).

I think that it is safe to accept it, especially as the anatomy is also not unfavourable to the cauline interpretation (fig. xxx, 5).

In its power of rapid elongation, the shoot of *Asparagus officinalis*, L., forms a sharp contrast to the slow-growing axes generally characteristic of Monocotyledons. One stem, growing out-of-doors near Cambridge, which I measured[1] on May 12, at the stage at which it was ready to cut for

Fig. xxx. *Asparagus* "needles." Figs. 1 and 2, *A. Sprengeri*, Reg. Figs. 1 A—E, transverse sections from series through leaf, *l.*, with its basal spine, *s.*, and the needles v, w, x, y, z, which it encloses; l'_1, l'_2, l'_3, the leaves associated with the needles w and x; w' and x' are rudiments which would have developed into two more needles (× 47). Figs. 2 A and B, two sections through a group of three needles, older than fig. 1 (× 47); the only leaves associated with the needles are l'_2 and l''_2. Fig. 3, *A. verticillatus*, L., flowering shoot (× ½). Figs. 4 and 5, *A. officinalis*, L. Figs. 4 A—C, transverse sections from a series through the needles v, w, x, y, z, developed in the axil of a leaf (× 47). Fig. 6, *A. plumosus*, Baker, fascicle of needles in axil of leaf (× 77). [A.A.]

market, had an above-ground portion which was 6·3 cms. long; in the course of the next week it had nearly trebled its length, reaching 18·3 cms., but it remained unbranched. In the second week, its length increased by more than 350°/₀, reaching a total of 84 cms.; it also produced a number of branches, the lowest of which was 18 cms. long—about the length, that is to say, of the main axis itself a week before.

[1] Arber, A. (1920[5]), p. 7.

CHAPTER IV

THE FOLIAGE-LEAF: DESCRIPTION

(i) *Phyllotaxis*

SINCE leaves are essentially members of the shoot, rather than independent entities, it may be well, before considering them in detail, to turn for a moment to the question of their arrangement on the stem. Although in many Monocotyledons the phyllotaxis is as well defined as is usually the case in Dicotyledons, in others there is a strong tendency to asymmetry, with the result that it is often impossible to reconcile the leaf-plan with any of the recognised schemes. The shoot of the Crown-imperial, *Fritillaria Imperialis*, L., drawn in fig. lxix, p. 93, is an instance of this. The phyllotaxis was highly irregular, the leaves being placed in segments of close and very crowded spirals, sometimes with gaps between these segments, while, in places, almost a whorled effect was produced[1]. In some garden plants of the White Lily, which I examined many years ago at Ethel Sargant's suggestion, I found that, though the divergence was generally $\frac{2}{5}$, $\frac{3}{8}$, or $\frac{5}{13}$, one could discover such angles as $\frac{1}{2}$, $\frac{3}{7}$ and $\frac{6}{17}$, and also leaf-successions which it was impossible to fit into any spiral.

In cases in which the phyllotaxis is regular, there is no hard-and-fast distinction between that of Dicotyledons and Monocotyledons, but it is noticeable that there is, in Monocotyledons, a widespread tendency to distichy[2], which, in Dicotyledons, is a relatively rare arrangement. Fig. xxxi, B, shows two-ranked leaves in an Orchid, *Calanthe Veitchii*, Hort. Cases of distichy are met with in the Helobieae, Glumiflorae, Spathiflorae, Liliiflorae, Farinosae and Microspermae—that is to say, in most of the cohorts belonging to the class. And moreover, in many families, this type of phyllotaxis is extremely common; in the Gramineae and Orchidaceae, for instance, it is the usual plan. This two-ranked arrangement sometimes leads to the development of the leaves and the axillary buds all in one plane, and thus induces a fan-like effect such as that shown in the seedling of *Triglochin maritima*, L., drawn in fig. xxxi, A. In other plants, such as *Pandanus*, the Screw-pine, and some species of *Gasteria*, distichy, associated with a spirally-twisted axis, gives the plant a startlingly decorative effect.

---

[1] On these "growth whorls" see Schoute, J. C. (1922).

[2] Arber, A. (1918²), p. 486.

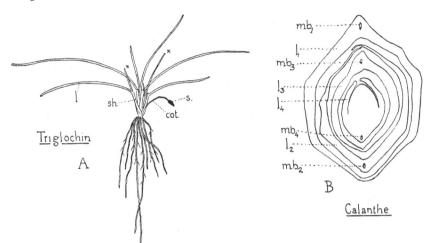

Fig. xxxi. Distichous phyllotaxis. A, *Triglochin maritima*, L., seedling (× ⅓); *s.*, seed; *cot.*, cotyledon; *sh.*, leaf-sheath; *l.*, leaf-limb; *x.*, lateral buds. B, *Calanthe Veitchii*, Hort., transverse section of apical bud, showing four of the inner leaves, l_1—l_4, with their midribs, $m.b_1$—$m.b_4$ (× 14). [A.A.]

(ii) *The Leaf-base*

The tendency to the development of an ensheathing leaf-base—though neither universal among Monocotyledons, nor confined to them—is widespread in the class, recurring in family after family. The sheath is often a prominent feature of the very young leaf, even if, at maturity, it is reduced to relative insignificance by the development of the limb. Figs. xxxii, 1 and 2, p. 56, illustrate this point for a species of *Narcissus*. In some instances the sheath may remain long in proportion to the limb, even when the leaf has completed its growth. This is strikingly exemplified in *Veratrum album*, L. (fig. xxxiii, p. 57). Here the mature vegetative plant seems, at first glance, to consist of an elongated axis, bearing a series of sessile leaves; in fact it has sometimes been so described. But closer inspection reveals the fact that the true stem is a small solid object hidden away at the base of the shoot (fig. xxxiii, B), while the apparent axis consists merely of leaf-sheaths. In the particular plant represented in fig. xxxiii, there were seven foliage leaves belonging to the current year, each of which had a sheath longer than the last, with the result that the part of the "axis" just below the lowest leaf-limb consisted of seven hollow cylindrical leaf-sheaths, fitted inside one another. The youngest and longest leaf-sheath was, at the base, enclosed inside the preceding six, but—leaving them behind one by one—it finally emerged alone at the top of the shoot; its length was 35 cms. The seven leaf-sheaths were slightly expanded at the base, and the innermost one enclosed a small terminal cone of tightly-packed leaves destined to unfold next year; the remains of some of last year's leaves were also present, forming a scaly coating to the base of the shoot.

The leaf-sheaths of *Veratrum* are very slightly succulent at the base, but, in many Monocotyledons, this succulence is much more pronounced, and the leaf-base region plays a prominent part in bulb formation. Sometimes, when the leaf-bases contain reserve material, the limb is eventually detached by means of an absciss layer; this occurs, for instance, in certain

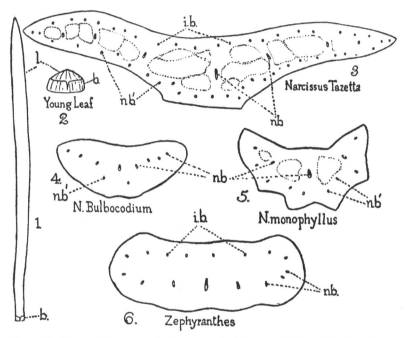

Fig. xxxii. Fig. 1, *Narcissus* sp. (garden var.): leaf showing relation of sheath to limb at maturity (× 0·5). Fig. 2, *Narcissus* sp. (garden var.): young leaf, slightly more than 1 mm. long, showing predominance of sheath. Fig. 3, *N. Tazetta*, L.: transverse section of limb of leaf (× 14). Fig. 4, *N. Bulbocodium*, L.: transverse section of limb of leaf (× 23). Fig. 5, *N. monophyllus*, T. Moore: transverse section of limb of leaf (× 23). Fig. 6, *Zephyranthes candida*, Herb.: transverse section of limb of leaf (× 14); *l.*, limb; *b.*, sheath; *n.b.* and *n.b'.*, series of normally orientated bundles; *i.b.*, series of inverted bundles (xylem in black, phloem in white, and outlines of lacunae in dotted line). [Arber, A. (1921³).]

Amaryllidaceae (*Narcissus, Galanthus* and *Leucojum*[1]). And in other plants, where bulbs are not formed, the leaf-bases may remain as a persistent armour, the upper part of the leaf being—as in the Amaryllids just mentioned—removed by means of a clean-cut absciss layer; this is particularly well seen in certain members of the Brazilian genus *Vellozia*, belonging to the family Velloziaceae, which is closely related to the Amaryllidaceae (fig. xxxiv).

[1] Parkin, J. (1898).

The sheathing region of the leaf may partially or completely encircle the stem, or, if it is open to the base, it may go even farther, forming a "wrap-over." Velenovský[1] has described a case from the Restionaceae—the scale-leaf of *Thamnochortus*—in which the spiral attachment executes more than two complete turns round the axis. A somewhat less extreme example

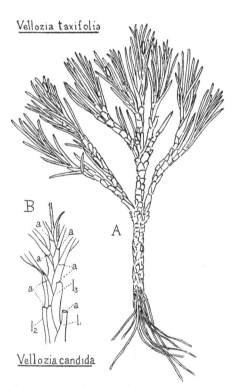

Fig. xxxiii. *Veratrum album*, L. A, vegetative plant at maturity, June 20, 1923 (× about ⅕). B, base of plant cut in two lengthwise (× about ⅖). [A.A.]

Fig. xxxiv. Persistent leaf-bases of *Vellozia*. A, *V. taxifolia*, Mart., habit drawing of plant (× ½); specimen from Brazil, Brit. Mus. Herbarium. B, *V. candida*, Mikan, part of shoot (× ½), Brit. Mus. Herbarium; in the leaves l_1, l_2, and l_3, the limb has become detached at the absciss layer, a, which is visible as a line in the younger leaves. [A.A.]

of "wrap-over" is seen in the spathe of *Arum maculatum*, L. (fig. cxlv, 2 C, p. 190).

The leaf-sheath of Monocotyledons is, indeed, so markedly liable to an exaggerated development, while that of the axis is correspondingly mini-

---

[1] Velenovský, J. (1907), fig. 358, p. 560.

mised, that we seem to be witnessing an actual transference of the focus of shoot-growth from axis to leaf-base. The result is that the shoot often appears, at first glance, to terminate in a leaf, rather than in a stem apex. Cases of this peculiarity may be found both among tuberous and elongated axes. As an example in which the stem is thickened and abbreviated, we may take the vegetative plants of *Sauromatum guttatum*, Schott, in which the first leaf of the season seems to arise as a terminal member from the centre of the top of the tuber; it is only on cutting through its base longitudinally that the truly terminal bud which it encloses is revealed (fig. clviii, 2, p. 208).

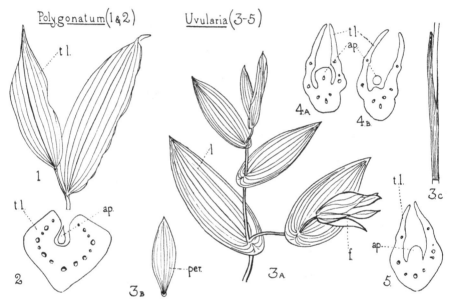

Fig. xxxv. "Terminal" leaves are throughout marked *t.l.* Figs. 1 and 2, *Polygonatum* sp.; fig. 1, apex of a flowering shoot which had no branches in the axils of the upper leaves ($\times \frac{1}{2}$); fig. 2, transverse section close to base of leaf *t.l.* in fig. 1, showing the vestigial stem apex, *ap.*, with no vascular tissue ($\times 14$). Figs. 3—5, *Uvularia* sp.; fig. 3 A, shoot ($\times \frac{1}{2}$); *f.*, flower, of which one perianth member, *per.*, with basal nectary pocket, is drawn in fig. 3 B ($\times \frac{1}{2}$); below the leaf *l.* in fig. 3 A, there was one additional normal foliage leaf, and then the leaf, consisting of a sheath alone, closed in the basal region, which is shown in fig. 3 C ($\times \frac{1}{2}$). Figs. 4 A and B, and 5, sections ($\times 23$) from the tips of two shoots to show apex, *ap.*; for further description see text. [A.A.]

In *Polygonatum*, Solomon's-seal, on the other hand, we find instances of a pseudo-terminal leaf borne by a slender elongated stem; fig. xxxv, 1, shows the end of a flowering shoot, in which, if one trusted to naked-eye observation, it might be supposed that the leaf *t.l.* actually terminated the axis. Microtome series through the critical region, however, reveal the fact that there is a vestigial stem apex (*ap.*, fig. xxxv, 2), which rises above the level of exsertion of the leaf, but is non-vascular, despite the fact that the shoot has reached complete maturity. *Uvularia* (figs. xxxv, 3—5) is even more striking in this respect, though the two shoots of an unidentified species, which I examined, did not show so extreme a state of things as that reported by Queva for *Uvularia grandiflora*, Sm., which he interprets as distinguished by the complete absence of a "cône végétatif terminal[1]." In one of the two shoots of which I cut microtome series (figs. xxxv, 4 A and B), the stem apex (*ap.*) rose as a minute cone above the exsertion of the last leaf; it was vascular below the detachment of the leaf, but non-vascular in its free region. In the second shoot (fig. xxxv, 5) the stem apex contained no vascular tissue and had no free region, but died out while still attached to the ventral surface of the leaf-base. This example represents the nearest approach, which I have myself seen, to a terminal position in the case of an organ which is indisputably a normal foliage leaf. We will consider later (pp. 136–148) the controversial case of the "phylloclades" of *Myrsiphyllum* and the Rusceae, which provide—on my view—examples of foliar members terminating "short shoots."

It is sometimes stated in text-books, that one of the characteristics distinguishing Monocotyledons is the lack of stipules—a statement which, however, needs some qualification. It is true that the leaf-base of most Monocotyledons is simple and sheathing (e.g. *Streptolirion*, fig. lxxxi, 9, p. 108), but there are, on the other hand, many cases, especially among water and marsh plants, in which this region is more highly elaborated. *Hydrocharis*, the Frog-bit (fig. xxxvi, 4, p. 60, and figs. xxxvii, 4 A—D, p. 61), may be cited as a plant in which there are two large and distinct stipules, while in *Potamogeton* (figs. xxxvi, 3 A and B; figs. xxxvii, 1 A and B) and *Calla* (figs. xxxvi, 2 A—C) there is a conspicuous ligular sheath, enclosing the younger leaves. Perhaps the most remarkable ligules among Monocotyledons are to be found in the Pontederiaceae. The ligular sheath of the Water-hyacinth, *Eichhornia speciosa*, Kth., is shown in figs. lxxxi, 8 A and B, p. 108, and that of the Pickerel-weed, *Pontederia cordata*, L., in fig. xxxvi, 1, p. 60; these ligules are trilobed at the apex, the terminal lobe suggesting a miniature lamina of membranous texture. But, more common than these cases of conspicuous free ligular sheaths, are those of slight upward prolongations of the sheath margin, to form a border or frill at the limit of sheath and limb. Such ligules are shown for *Allium* in figs. lxxvi, 22 A, 23 A, 25 A, p. 101, and for the Gramineae in figs. cv, p. 134, cxxxiv, p. 167, and cxxxv, p. 169. The relations which the various leaf-base modifications met with in Monocotyledons bear to one another will be considered later (p. 96).

[1] Queva, C. (1907)

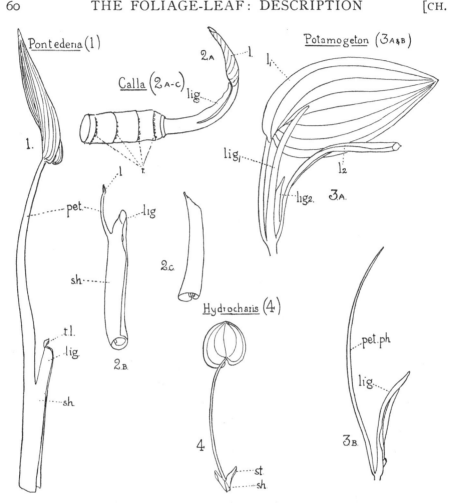

Fig. xxxvi. Ligules. Fig. 1, *Pontederia cordata*, L., leaf showing sheath, *sh.*, and ligule, *lig.*, with terminal lobe, *t.l.* ($\times\frac{1}{2}$). Figs. 2 A—C, *Calla palustris*, L. ($\times$ 1); fig. 2 A, apex of rhizome, all but youngest leaves dissected away; *lig.*, ligule of leaf *l.*, enclosing terminal bud; *r.*, root rudiments at each node. Figs. 2 B and C, reduced leaves from the base of an axillary shoot; fig. 2 B, petiole shortened and blade reduced to a tiny rudiment, *l.*; fig. 2 C, blade absent, petiole reduced to a mere point; ligule absent. Figs. 3 A and B, *Potamogeton natans*, L. ($\times\frac{1}{2}$); fig. 3 A, end of shoot; leaf l_1 with ligule, *lig.*$_1$, and leaf l_2 with ligule, *lig.*$_2$. Fig. 3 B, petiolar phyllode, *pet.ph.*, with ligule, *lig.* Fig. 4, *Hydrocharis Morsus-ranae*, L., a small leaf ($\times\frac{1}{2}$); leaf-sheath, *sh.*, with two stipular wings, *st.* [A.A.]

(iii)   *The Leaf-limb*

Whereas the Dicotyledonous leaf has, as a rule, both a petiole and an expanded lamina, that of the Monocotyledon very frequently has a limb of a simple linear type, with no differentiation into stalk and blade. Such limbs are sometimes approximately cylindrical, with a more or less radial type of anatomy (e.g. certain species of Onion, *Allium*, fig. lxxvi, p. 101, Arrow-

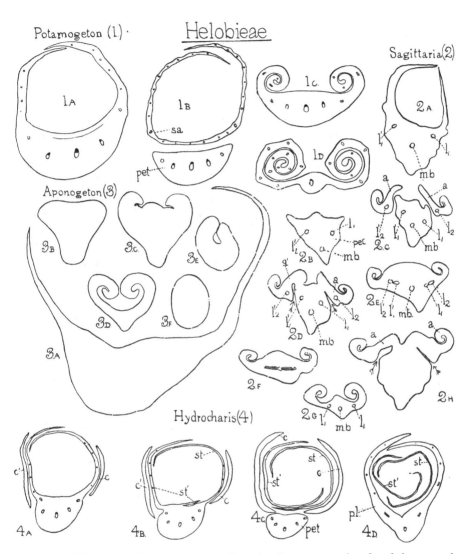

Fig. xxxvii. Figs. 1 A—D, *Potamogeton natans*, L., series of transverse sections from below upwards through a young leaf, from sheath, fig. 1 A, to "lamina," fig. 1 D ($\times 47$); *pet.*, petiole; *s.a.*, *stipula adnata*. Fig. 2, *Sagittaria sagittifolia*, L., figs. 2 A—G, serial sections through one young leaf ($\times 23$) from sheath, fig. 2 A, to "lamina," fig. 2 G; *a.* and *a'.*, auricles; *pet.*, petiole; *m.b.*, median bundle; l_1 and l'_1, main lateral bundles; l_2 and l'_2, bundles given off from l_1 and l'_1 to supply auricles; fig. 2 H, transverse section of another young leaf ($\times 23$) at level of detachment of auricles, vascular bundles omitted. In figs. 2 D and 2 H the arrows indicate the invaginations which detach the auricles. Figs. 3 A—F, *Aponogeton distachyum*, Thunb., series of transverse sections through one leaf from below upwards ($\times 47$), vascular bundles omitted. Figs. 4 A—D, *Hydrocharis Morsus-ranae*, L., series of transverse sections through one leaf passing from sheath, fig. 4 A, to "lamina," *p.l.*, fig. 4 D; *pet.*, petiole; *st.* and *st'.*, stipules; *c.* and *c'.*, basal cordate lobes of "lamina" ($\times 23$). [Arber, A. (1922[4]).]

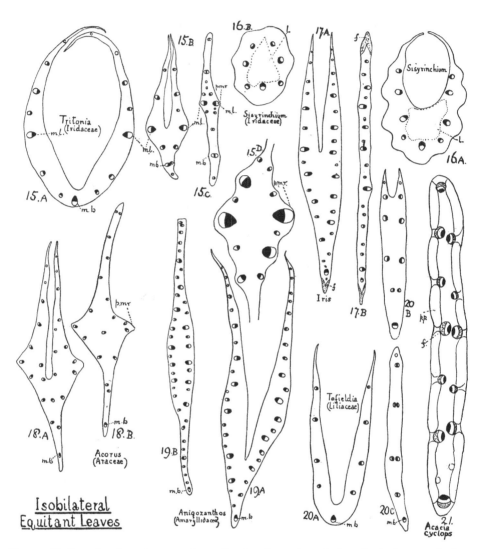

Fig. xxxviii. Lettering throughout:—*m.b.*, median bundle; *m.l.*, main lateral; *l.*, lacuna; *f.*, fibres; *p.p.*, palisade parenchyma. Xylem is represented solid black, phloem white, and fibres dotted. Figs. 15 A—D. *Tritonia* (garden hybrid). Figs. 15 A, B, C. Successive transverse sections of young leaf, passing through basal sheathing, and upper regions. Fig. 15 D. The pseudo-midrib (*p.m.r.*) from an older leaf (× 23). Figs. 16 A and B. Transverse sections of sheathing and upper regions of leaf of *Sisyrinchium* sp. (× 23 *circa*). Figs. 17 A and B. *Iris* sp. Transverse sections of sheathing and upper regions of leaf (× 7). Figs. 18 A and B. *Acorus Calamus*, L. Transverse sections of sheathing and upper regions of leaf (× 7), *p.m.r.*, pseudo-midrib. A series of very small bundles lying close to both surfaces between the larger bundles has, for simplicity, been omitted. Figs. 19 A and B. *Anigozanthos* sp. Transverse sections of sheathing and upper regions of leaf (× 7). Figs. 20 A—C. *Tofieldia calyculata*, Wahl. Series of transverse sections through basal sheathing and upper regions of leaf (× 23). Fibrous bundle-sheaths not represented. Fig. 21. Transverse section of phyllode of *Acacia cyclops*, A. Cunn. (× 23). Some of the smaller bundles omitted. [Arber, A. (1918²).]

grass, *Triglochin*, fig. lxxxiii, 10, p. 110, and *Sisyrinchium*, fig. xxxviii, 16 B). They may, on the other hand, be flattened in the vertical plane, giving the so-called isobilateral equitant leaf, examples of which are shown in fig. xxxviii. Others again are flattened in the horizontal plane, with the vascular strands—or at least the minor ones—forming a peripheral series, with the xylem pointing inwards (e.g. *Brodiaea*, fig. xxxix; *Eremurus*, fig. lxxxiii, 6, p. 110, and certain species of *Allium*, fig. lxxvi, 26 B, etc., p. 101). Besides these limbs, which show some indication of radial symmetry, and others, derivable from the radial type, but of a more complex form, which

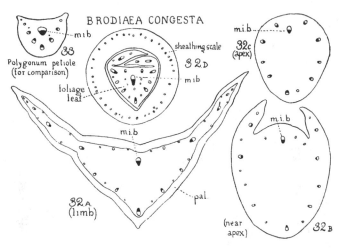

Fig. xxxix. Figs. 32 A—D, *Brodiaea congesta*, Sm.; *m.i.b.*, main inverted bundle. Fig. 32 A, transverse section of limb of leaf; *pal.*, palisade parenchyma (× 11). Fig. 32 B, transverse section of another leaf near apex (× 18). Fig. 32 C, transverse section close to extreme apex (× 18). Fig. 32 D, base of sheathing leaf and first foliage leaf of young vegetative shoot (× 11). Fig. 33, *Polygonum amphibium*, L., transverse section of petiole, for comparison with limb of *Brodiaea*; *m.i.b.*, main inverted bundle (× 11). [Arber, A. (1920³).]

we shall discuss later (p. 117 *et seq.*), there is another fairly distinct type known as the ribbon-limb, which is entirely dorsiventral, and possesses only one series of bundles; it occurs in such plants as the Grasses (fig. cxxxiv, p. 167) and certain Cyperaceae, e.g. the Bulrush, *Scirpus lacustris*, L. (fig. xl, p. 64), and other aquatics (figs. ciii, p. 129, cxli, p. 184, cxlii, p. 185).

In addition to the leaves so far considered, in which there is no distinction between stalk and blade, there are other Monocotyledonous types, which possess an expanded lamina, sharply marked off from a stalk-like region. It is rather surprising, that, although Monocotyledons are frequently bladeless, the blades, when they occur, sometimes—e.g. in certain Palms— attain colossal proportions. Fig. xli, p. 64, shows a simple type of ovate blade which recurs again and again among Monocotyledons. One highly characteristic modification of this type is that in which the base is either cordate or sagittate[1]; examples of this form are met with in the Alismaceae (fig. xlii); Butomaceae; Hydrocharitaceae (fig. xxxvi, 4, p. 60); Pontederiaceae (fig. lxxxi, 7, p. 108); Cyanastraceae; Commelinaceae (fig. lxxxi, 9, p. 108); Roxburghiaceae; Liliaceae; Amaryllidaceae (fig. c, p. 126); Dioscoreaceae (frontispiece and fig. ci, p. 127); Araceae (fig. cxlv, p. 190); Gra-

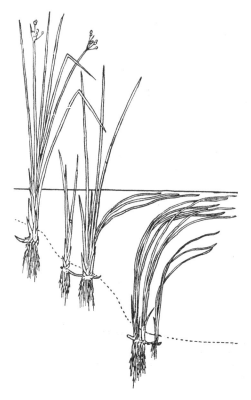

Fig. xl. Great Sedge (*Scirpus lacustris*, L.), showing plants in deep water with ribbon-leaves only; in shallower water with ribbon-leaves and flowerless scapes; and in very shallow water without ribbon-leaves and with both flowerless and flowering scapes; river bed seen in section. [Lister, G. (1920).]

Fig. xli. *Funkia grandiflora*, Sieb. et Zucc. Upper part of petiole and "lamina" to show venation (reduced). [Arber, A. (1918[2]).]

mineae[2]; and Orchidaceae (fig. clx, 8, p. 229). In these cordate and sagittate leaves, the venation of the basal lobes is not always referable to a single scheme, but conforms to one of two distinct types. Either the marginal veins of the blade curve downwards and outwards, following the outline of the lamina (e.g. *Alisma parnassifolium*, Bassi, var. *majus*, fig. lxxxiv, 6, p. 111,

[1] Arber, A. (1918[2]), p. 471; (1920[2]), p. 441; (1922[5]), p. 89.
[2] E.g. *Phyllorhachis sagittata*, Trim.

and *Streptolirion volubile*, Edgw., fig. lxxxi, 9, p. 108), or else the two main lateral veins, and some of the minor ones, are in no part of their course directed towards the leaf apex, but pass immediately into the basal lobes, where they terminate without recurving. This latter type of venation, which we may distinguish as the sagittate type proper, is shown in figs. xlii, A—E, below. It is possible that such schemes of venation as those represented in

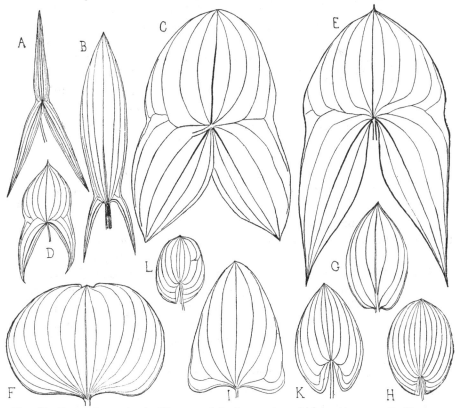

Fig. xlii. Types of venation in Alismaceae (all × ¼, except A, which is × ⅛). A—E, *Sagittaria*; A, *S. sagittifolia*, L., from Burmah; B, *S. macrophylla*, Zucc.; C, *S.* sp. (garden var.); D, *S. latifolia*, Willd.; E, *S.* sp., from Java. F—H, *Alisma*; F, *A. parnassifolium*, Bassi, var. *majus*; G, *A. Plantago*, L., var. *americanum*, Gray; H, *A. acanthocarpum*, F. Mueller. I, *Echinodorus radicans*, Engelm. K and L, leaves from unidentified specimens. [A.A.]

figs. xlii, H and K, and fig. lxxxiv, 8, p. 111, may be regarded as intermediate links, but I am inclined to think that they are merely a modification of the first, or cordate type, and do not lead on to the true sagittate type.

The cordate leaves of certain Dioscoreaceae are characterised by the precocious development of the leaf apex to form the so-called "forerunner tip" (frontispiece; fig. xcvii, 20, p. 123; fig. ci, A, *M*, p. 127). In *Dioscorea macroura*, Harms., these forerunner tips bear glands, which are described as affording a home to a nitrogen-fixing bacterium[1].

[1] Orr, M. Y. (1923).

The perfoliate leaf of *Uvularia* (fig. xxxv, 3 A, p. 58) may be treated as a development of the cordate type, in which the basal lobes of the blade are united and encircle the axis. That the perfoliation is a secondary development is indicated by the fact that the leaves do not show it in their youngest stages (figs. xxxv, 4 and 5).

An unusual modification, met with in the limbs of certain Monocotyledonous leaves, is a crisping or undulation of the margin, which gives an

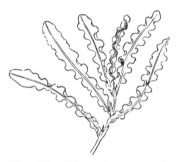

elegant frilled appearance. This character is most frequent in aquatics, but is not confined to them. The following list of plants with crisped leaf-margins is, no doubt, far from complete, but it includes all the cases with which I am acquainted: *Aponogeton ulvaceus*, Baker; various species of *Potamogeton*, e.g. *P. crispus*, L. (fig. xliii); *Echinodorus Martii*, Mich. (*Alisma intermedium*, Mart.)[1]; *Eucomis undulata*, Ait.; *Crinum natans*, R.Br.; *Tritonia undulata*, Baker (*Ixia crispa*, L.)[2]; *Tritonia viridis*, Ker-Gawl.

Fig. xliii. *Potamogeton crispus*, L., end of shoot ($\times \frac{1}{2}$). [A.A.]

Riede[3], who has made a study of this peculiar leaf form with special reference to *Aponogeton ulvaceus*, shows that it is due to the existence of different rates of growth in the different regions of the limb. He notes that in a leaf whose median length was 20 cms., the length of the margin was 60 cms.

A rare form of Monocotyledonous leaf is that in which the margin, instead of being sinuous, as in the species just considered, is rolled downwards, e.g. *Philesia magellanica*, G. F. Gmel. (fig. lxvii, 7, p. 90). An upward curvature of the margins, such as that in *Aphyllanthes* (fig. lxvii, 6, p. 90) is less unusual.

A few Monocotyledonous leaves have the peculiarity of developing a basal twist, so that the morphologically lower (dorsal) surface is turned towards the sky. One of our British plants, Ramsons, *Allium ursinum*, L., regularly behaves in this way (fig. xliv, 28 A—E). The twist here begins in the upper part of the sheath; the petiole at the level Y is completely inverted, and from the base of the limb (fig. 28 E) upwards, the xylem of the bundles is directed towards the lower surface. Since the leaf-bearing axis of *A. ursinum* is entirely subterranean, and the torsion of the leaves takes place below the level of the ground, the fact of their inversion is not obvious in the field. But there are other Monocotyledons in which the twisting of the leaf-bases cannot be overlooked, since the stems are aerial. Two cases from the Amaryllidaceae—*Bomarea* and *Alstroemeria*—

[1] Seubert, M., in Martius, C. F. P. von (1840, etc.), vol. iii, p. i, pl. 14.
[2] Curtis's *Bot. Mag.* t. 599.   [3] Riede, W. (1920).

ALLIUM (sections—MOLIUM, NECTAROSCORDUM MICROSCORDUM)

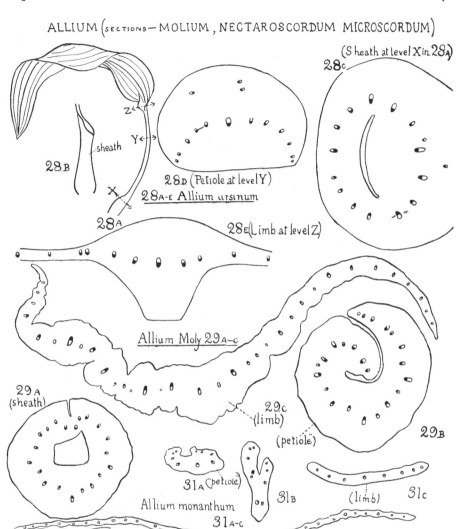

Fig. xliv. Leaf structure of *Allium*. Figs. 28 A—E, *Allium ursinum*, L. (Sect. *Molium*). Fig. 28 A, leaf of non-flowering plant (½ nat. size); fig. 28 B, sheathing base of leaf of flowering plant (½ nat. size); fig. 28 C, transverse section of sheath at level X; fig. 28 D, transverse section of petiole at level Y; fig. 28 E, transverse section of limb at level Z (figs. 28 C, D, E, × 14). (Note partial twisting in fig. 28 C, and inversion in figs. 28 D and E.) Figs. 29 A—C, *Allium Moly*, L. (Sect. *Molium*). Fig. 29 A, sheath just above the great swelling which forms the bulb; fig. 29 B, apparent petiole; fig. 29 C, limb of leaf. (All × 14.) Figs. 30 A and B, *Allium Dioscoridis*, Sibth. et Sm. (Sect. *Nectaroscordum*). Fig. 30 A, transverse section of limb of leaf (× 14), *k.*, keel; fig. 30 B, margin of limb in fig. 30 A, further enlarged to show orientation of marginal bundles (× 47). Figs. 31 A—C, *Allium monanthum*, Maxim. Fig. 31 A, transverse section of apparent petiole; fig. 31 B, transverse section of intermediate region; fig. 31 C, transverse section of base of limb. (All × 14.) There is probably a good deal of distortion in figs. 30 and 31, due to imperfect recovery of form of the herbarium material used. [Arber, A. (1920³).]

are shown in fig. xlv; for clearness, the morphologically upper surfaces of all the leaves are shown in black, the under surfaces being left white. It will be noticed that, in the shoot of *Bomarea* drawn in fig. xlv, 1, the youngest leaves are held in a normal attitude, but torsion of the leaf-base has occurred

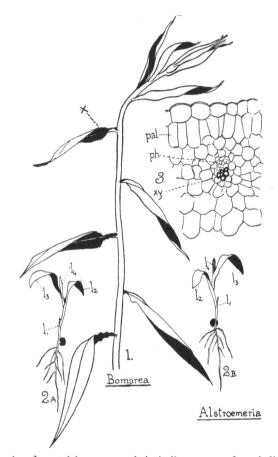

Fig. xlv. Inverted leaves; morphologically upper surfaces indicated in black. 1, *Bomarea cantabrigiensis*, Lynch (hybrid between *B. Caldasiana*, Herb., and *B. hirtella*, Herb.). Apex of shoot ($\times \frac{1}{2}$). 2 and 3, *Alstroemeria*. 2 A and B, one seedling of *A.* sp., viewed from two sides ($\times \frac{1}{2}$). 3, *Alstroemeria aurantiaca*, D.Don, small part of transverse section of leaf, to show palisade parenchyma (*pal.*) towards upper (morphologically lower) surface, and xylem (*xy.*) directed towards lower (morphologically upper) surface ($\times 193$). [A.A.]

in that marked x, and is characteristic of all the older leaves. Two views of a seedling of *Alstroemeria*, the Peruvian Lily, are drawn in figs. xlv, 2 A and B. The youngest leaf, l_4, is vertical, but the leaves l_2 and l_3 are twisted. Sections of the resupinated leaves of these Amaryllids have a strikingly upside-down appearance, for while there is palisade tissue towards

the upper (morphologically lower) surface, the bundle has its xylem directed
towards the surface which is actually the lower (fig. xlv, 3). A number of

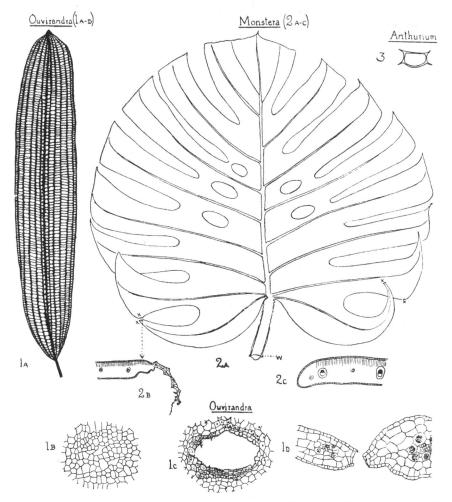

Fig. xlvi. Fenestration. Figs. 1 A—D, *Ouvirandra fenestralis*, Poir. Fig. 1 A, leaf-limb (reduced) [A.A.]. Figs. 1 B—D, details of fenestration [Sergueéff, M. (1907)]. Fig. 1 B,
suberised zone around the future perforation; fig. 1 C, adult perforation, seen in tangential section;
fig. 1 D, transverse section of perforation. Figs. 2 A—C, *Monstera deliciosa*, Liebm. Fig. 2 A, young
leaf (much reduced); *w.*, winged margin of petiole; at the points marked with crosses on either
side, detachment of the basal lobe has just taken place; fig. 2 B, transverse section through margin
of detached lobe; the dotted part to the right is shrivelled and dead (× 21); fig. 2 C, transverse
section of margin of one of the oval perforations to show completion of epidermis (× 21). Fig. 3,
Anthurium magnificum, Lind., petiole cut across to show winging. [A.A.]

examples of leaves which are more or less completely inverted are found in
the Gramineae (e.g. *Pharus*)[1].

[1] Lindman, C. A. M. (1899).

The majority of Monocotyledonous leaves—including all those to which we have hitherto referred—are of a simple type, with a more or less entire margin. But others are compound, conforming either to the pinnate or the palmate type. Figs. ci, E and F, p. 127, show the palmate compound leaf of

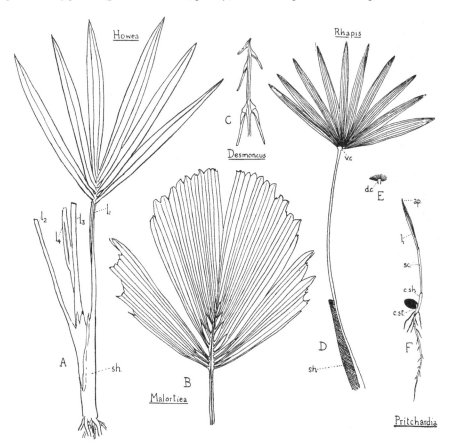

Fig. xlvii. Palm leaves. A, *Howea Forsteriana*, Becc., young plant (reduced) to show form of leaf, l_1; leaves l_2 and l_3 cut off short, and others removed; *sh.*, leaf-sheath. B, *Malortiea gracilis*, Wendl., leaf (× rather less than ½) to show fenestration; specimen from Mexico, Kew Herbarium. C, *Desmoncus* sp., apex of leaf to show reduced pinnules forming recurved spines (× rather less than ¼); specimen from Trinidad, British Museum Herbarium. D and E, *Rhapis humilis*, Blume; D, leaf (× rather less than ¼); *v.c.*, ventral crest; E, back view of base of limb, to show *d.c.*, dorsal crest. F, *Pritchardia filifera*, Linden, seedling (× rather less than ½); *c.st.*, cotyledon stalk; *c.sh.*, ligular sheath of cotyledon; *sc.*, first scale-leaf; l_1, first foliage leaf; *ap.*, solid apex of l_1. [A.A.]

Dioscorea pentaphylla. In the case of certain Aroids, the pinnules of the pinnate leaf are individually deciduous[1]. In addition to these typically compound leaves, which approximate in appearance to those of Dicotyledons, there are certain complicated forms of leaf-limb which are peculiar

[1] Engler, A. (1881—1884).

to Monocotyledons. One of these is found in *Ouvirandra fenestralis*, Poir., the Water-yam, or Lattice-leaf of Madagascar, in which the blade at maturity consists merely of a network with rectangular meshes (fig. xlvi,

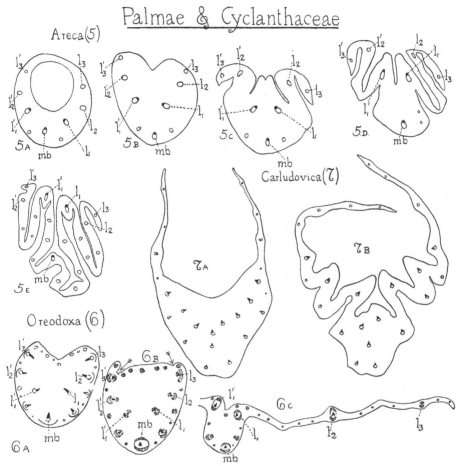

Fig. xlviii. Figs. 5 A—E, *Areca sapida*, Soland., series of transverse sections (× 47) through one leaf, from sheath, fig. 5 A, to plicate "lamina," fig. 5 E; *m.b.*, median bundle; l_1, l_2, l_3, l'_1, l'_2, l'_3, principal lateral bundles. In figs. 5 C—E the invaginations are shown which penetrate between the bundles of the petiole, fig. 5 B. Figs. 6 A—C, *Oreodoxa regia*, H. B. et K., sections from a series through the first foliage leaf (third plumular leaf) of a seedling (× 14). Lettering of bundles as in fig. 5; fig. 6 A, petiole; fig. 6 B, first signs of invaginations indicated by arrows; fig. 6 C, half the "lamina," showing eventual distribution of bundles. Figs. 7 A, B, *Carludovica Plumerii*, Kunth, transverse sections of young leaf passing through sheath, fig. 7 A, and region where invagination begins, fig. 7 B (× 14). [Arber, A. (1922[4]).]

I A, p. 69). The pattern produced owes its formal geometrical character to the Monocotyledonous venation of the leaf. The limb in its youngest stages is imperforate, and Serguéeff[1] has shown that the position of the

Serguéeff, M. (1907).

future mesh is first marked out by the suberisation of a ring or rectangle of cells belonging to the mesophyll between the veins (fig. xlvi, I B, p. 69). The elements isolated in this manner die and drop away, so that finally a hole is left surrounded by a corky boundary (fig. xlvi, I C). A somewhat different

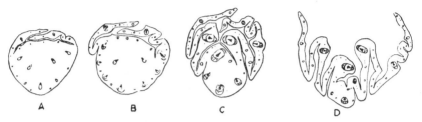

Fig. xlix. *Oreodoxa regia*, H. B. et K. Series of transverse sections through upper part of petiole, and limb, of third plumular leaf (first foliage leaf, which was preceded by two scale-leaves) of seedling (× 14) ; A, early stage of invagination of petiole. [Arber, A. (1922²).]

form of fenestration, also due to necrosis of a defined area of tissue, is charac-teristic of certain Aroids, such as *Monstera deliciosa*, Liebm. (fig. xlvi, 2 A—C). In this plant, however, the perforations have a more "finished" appearance than in *Ouvirandra*, for transverse sections of the margin of one of the oval holes show that, in *Monstera*, there is no disjunction between the upper and lower epidermis, which become, to all appearance continuous (cf. figs.

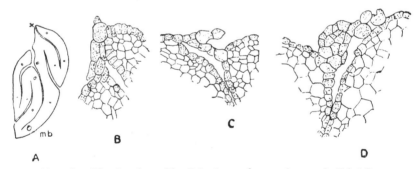

Fig. 1. Epidermal proliferations in seedling Palm leaves, from sections cut by Ethel Sargant. A—C, *Thrinax excelsa*, Lodd. A, transverse section of third plumular leaf (× 47) ; *m.b.*, median bundle. B, mouth of invagination marked × in A (× 318). The epidermis and its proliferation, which occludes the opening of the groove, are dotted. C, another occluded groove, from lower down in the same leaf (× 318). D, *Trachycarpus Fortunei*, H. Wendl. Mouth of invagination occluded by epidermal outgrowth, from a transverse section similar to that shown in fig. xc, F, p. 118 (× 318). [Arber, A. (1922²).]

xlvi, 2 C and I D). The perforations may eventually be opened up as far as the leaf edge, by a slit crossing the marginal tissues. Such splitting has recently taken place in the case of the two basal perforations marked with a × in fig. xlvi, 2 A. Fig. xlvi, 2 B, which is a transverse section across the margin of the slit, shows that this detachment has been brought about through the death of the shrivelled tissue on the right-hand side.

Fenestration may also occur in the Palms; the case of *Malortiea gracilis*, Wendl. (*Geonoma fenestrata*, Wendl.) is illustrated in fig. xlvii, B, p. 70. But, in this family, a more important result than fenestration is brought about by necrosis and splitting: for these factors are also responsible for the "compound" character of the elaborate "fan" and "feather" blades characteristic of the Palms. Transverse sections of the young leaf-rudiments (e.g. fig. xlviii, 5 E, p. 71, and fig. xlix, D) show a form which at first sight suggests an origin by folding. We will consider the meaning of this form later (p. 118); it will suffice, for the moment, merely to note its character. The next stage is a curious one, and, so far as I know, unparalleled in other groups. The "folds" become cemented together by the epidermal proliferations shown in figs. l, B, C and D. At a still later stage, by mucilaginous degeneration of

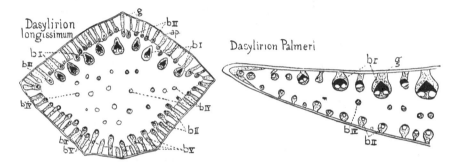

Fig. li. *Dasylirion* (xylem, black; phloem, white; fibres, dotted). Left-hand figure, transverse section of limb of leaf of *D. longissimum*, Lem. (× 12½, circa); *b* 1—*b* v, bundles belonging to different series; *g.*, fibrous girder; *a.p.*, assimilating parenchyma. Right-hand figure, transverse section of leaf-margin of *D. Palmeri*, Trelease (× 12½, circa). [Arber, A. (1920[3]).]

meristematic cells, or by drying and tearing along certain lines[1], the leaf is split into segments which produce the complex pinnuled form characteristic of most Palm leaves at maturity (fig. xlvii, p. 70). The interpretation of the so-called "ligule," which occurs at the base of the limb in the Fan-palms, will be discussed on p. 119. A few climbing Palms show the farther modification that some of the terminal pinnules are solid instead of leafy, and take the form of recurved spines (e.g. *Desmoncus* sp., fig. xlvii, C, p. 70).

(iv) *Anatomy*

An anatomical feature in which Monocotyledonous leaves often differ from those of the average Dicotyledon, is the poor differentiation of the assimilating tissue. It is true that there are many examples of Monocotyledonous leaves in which the palisade tissue is sharply distinguished from

[1] Deinega, V. (1898).

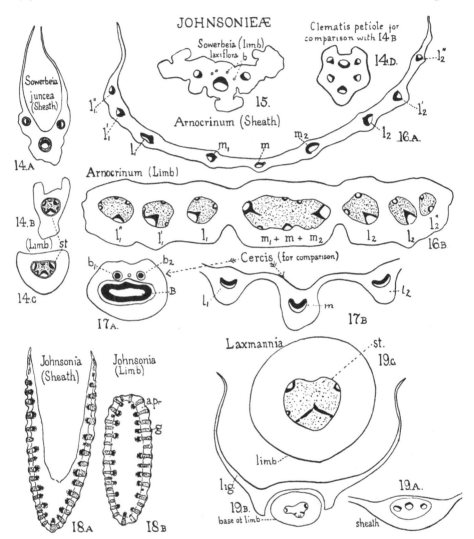

Fig. lii. Leaf structure of Johnsonieae. (Xylem, black; phloem, white; fibres, dotted.) Figs. 14 A—C, *Sowerbeia juncea*, Sm. Fig. 14 A, transverse section through leaf-sheath. Figs. 14 B and C, transverse sections through limb; *st*, axial bundle-group. (All × 14.) Fig. 14 D, *Clematis Vitalba*, L. Transverse section of petiole for comparison with limb of *Sowerbeia juncea*, Fig. 14 C (× 14). Fig. 15, *Sowerbeia laxiflora*, Lindl. Transverse section of limb (× 23); *b.*, small additional bundles. Figs. 16 A and B, *Arnocrinum Drummondii*, Endl. Fig. 16 A, transverse section of leaf-sheath (× 14); $l_1, l'_1, l''_1, l_2, l'_2, l''_2, m_1, m_2$, lateral bundles; *m.*, midrib. Fig. 16 B, transverse section of limb of another leaf (× 14); lettering corresponds to fig. 16 A (see fig. liii, p. 75, for one bundle-group on a larger scale). Figs. 17 A and B, *Cercis Siliquastrum*, L., for comparison with *Arnocrinum*. Fig. 17 A, transverse section of petiole, showing steles B, b_1, and b_2 (× 14). Fig. 17 B, transverse section of base of lamina including midrib, *m.*, and two main laterals, l_1 and l_2 (× 14). Figs. 18 A and B, *Johnsonia lupulina*, R.Br. Fig. 18 A, transverse section of leaf-sheath (× 14); *g.*, fibrous girder; *a.p.*, assimilating parenchyma. Figs. 19 A—C, *Laxmannia grandiflora*, Lindl. Figs. 19 A and B, transverse sections in sheath region (× 14); *lig.* = ligule. In fig. 19 A the wings of the sheath are omitted. Fig. 19 C, transverse section of limb (× 23). For a more highly magnified drawing of the central bundle-group (*st.*) see fig. liii. [Arber, A. (1920³).]

the spongy parenchyma. The floating leaf of *Trianea bogotensis*, Karst. (fig. xcix, p. 125), may be taken as an instance; here the spongy parenchyma is developed in an exaggerated form as an air-tissue. But in a very large number of Monocotyledonous leaves, the whole mesophyll consists of roundish or irregular cells, and it is scarcely possible to distinguish a palisade region. The idea suggests itself that this lack of morphological differentiation may perhaps be in some way correlated with that chemical difference between the two classes, which brings many Monocotyledons into the sugar-leaved category, while Dicotyledons are prevailingly starch-leaved.

The main feature distinguishing the general skeletal system of Monocotyledonous leaves from that of Dicotyledons is the frequency of a type

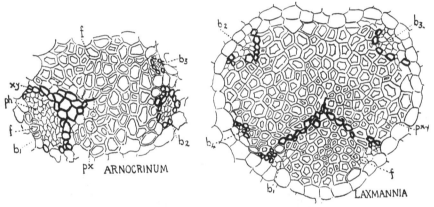

Fig. liii. Johnsonieae. *Arnocrinum Drummondii*, Endl. A lateral bundle-group from a section similar to that drawn in fig. lii, 16 B, p. 74. It shows a group similar to l_1, including the three bundles, b_1, b_2, and b_3, embedded in fibres, *f.*; *xy.*, xylem; *ph.*, phloem; *px.*, protoxylem (× 250, *circa*). *Laxmannia grandiflora*, Lindl. Transverse section of central strand (*st.*) of leaf shown in fig. lii, 19 C, p. 74 (× 150, *circa*). [Arber, A. (1920³).]

in which the strands are arranged more or less radially, recalling the scheme met with in many Dicotyledonous petioles. Another common type is the simple, more or less linear, leaf with a single series of bundles. In either type various complexities may be introduced, such as the development of additional sets of strands, giving a multiseriate structure, e.g. *Dasylirion* (Liliaceae, fig. li, p. 73). Or again, it may occasionally happen that the single strands are replaced by more or less radial groups of bundles, enclosed in a common fibrous sheath. Such leaves may be called polystelic—using the term in a purely descriptive sense. *Arnocrinum Drummondii*, Endl., belonging to the Liliaceous tribe Johnsonieae, is a striking example (fig. lii, 16 B). In the leaf-limb of this plant I have seen seven steles, each including two to six bundles. One of the smaller of these steles is shown in greater detail in fig. liii, left-hand diagram. In the leaves of the genus *Xyris* (Xyridaceae), bundle-groups embedded in fibres, comparable with those of the Johnsonieae,

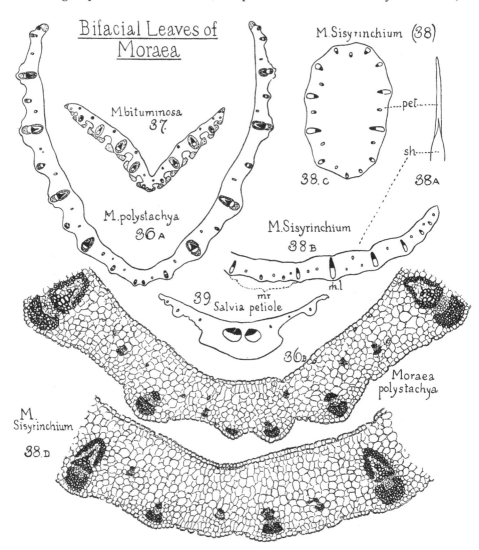

Fig. liv. Fig. 36, *Moraea polystachya*, Ker-Gawl.; fig. 36 A, transverse section of leaf (× 14); fig. 36 B, transverse section of midrib region of another leaf (× 47). Fig. 37, *M. bituminosa*, Ker-Gawl., transverse section of leaf (× 14). Fig. 38, *M. Sisyrinchium*, Ker-Gawl.; fig. 38 A, leaf apex seen from the ventral side (× ½) to show solid apex, *pet.*, and bifacial region, *sh.*; fig. 38 B, transverse section (incomplete) of bifacial region of leaf (× 14); *m.r.*, midrib region; *m.l.*, main lateral; fig. 38 c, transverse section of apical limb of leaf, *pet.*, in fig. 38 A (× 14); fig. 38 D, transverse section of midrib region indicated by bracket in fig. 38 B (× 47). Fig. 39, *Salvia Verbenaca*, L. (Labiatae), transverse section of petiole (× 14) to show absence of median bundle. [Arber, A. (1921¹).]

are to be observed. Figs. lxvi, 13 A—C, p. 88, give an idea of the complexity which these groups may reach.

A minor peculiarity of the general vascular scheme, which is common to a certain number of Monocotyledons, is the relative unimportance, or even absence, of the median bundle. This is very strikingly shown in the genus *Moraea* (Iridaceae)[1]. In *Moraea Sisyrinchium*, Ker-Gawl., for instance, the leaf-base forms the greater part of the leaf, and in this region, although the lateral bundles are well-developed, the part where one would expect to find the midrib (*m.r.*, figs. liv, 38 B and D) is occupied by several small bundles. The leaf terminates in a solid apex (*pet.*, fig. 38 A), and here, also, the median region has the same unusual character (fig. 38 C). *Moraea bituminosa*, Ker-Gawl. (fig. liv, 37), and *M. polystachya*, Ker-Gawl.

(figs. liv, 36 A and B) also have a poorer vascular supply in the "midrib" than in the rest of the leaf. Many other examples of the same peculiarity might be mentioned, but it will perhaps suffice to refer to the leaf-limbs of the genus *Astelia* (fig. lv, lower figure) and to the leaf-sheaths of some of the Palms, where this feature has been described[2].

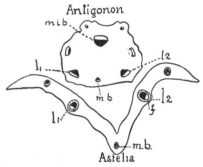

In certain Monocotyledons, the midrib region, instead of being marked by smaller bundles, as in the instances just cited, is indistinguishable from the rest of the leaf, and one can scarcely speak of symmetry about a midrib. This is the case, for example, in the main part of the leaf of *Anemarrhena*

Fig. lv. (Below) *Astelia Solandri*, A. Cunn. Transverse section of limb of leaf (× 23); *m.b.*, median bundle; l_1 and l_2, laterals; *f.*, fibres. (Above) *Antigonon leptopus*, Hook. et Arn. (Polygonaceae). Transverse section of petiole for comparison with *Astelia*, showing small size of main bundle (*m.b.*) in comparison with the laterals (l_1 and l_2); *m.i.b.*, main inverted bundle. (× 14.) [Arber, A. (1920[3]).]

(fig. lvi, 1, p. 78), although the median strand becomes conspicuous at the extreme apex. It must not be assumed, however, that the lack of a midrib is at all a general character of Monocotyledons; in the great majority of cases, the leaf is symmetrical about a midrib which is clearly defined, and both it and the main lateral veins are well supplied with vascular tissue (e.g. *Sagittaria*, fig. lvii, p. 78). In other instances, again, the midrib is conspicuous, but there is a curious lack of foliar symmetry. In certain Rapateaceae (e.g. *Cephalostemon affinis*, Koern., fig. lviii, 26 A, p. 79, and *Rapatea angustifolia*, Spr., fig. lviii, 25 A) the sheath is sharply folded, while the tissues thin markedly to the fold, which lies at a considerable distance from the median bundle.

The individual vascular strands of Monocotyledonous leaves are nearly always collateral, but they may occasionally become amphivasal, through the xylem elements creeping, as it were, around the phloem, until they

[1] Arber, A. (1921[1]), p. 328.   [2] Schoute, J. C. (1915).

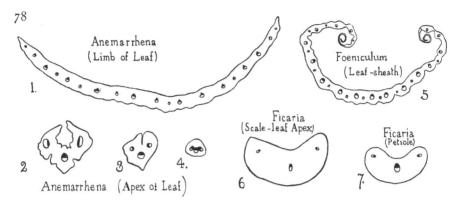

Fig. lvi. (Xylem, black; phloem, white.) Figs. 1—4, *Anemarrhena asphodeloides*, Bunge. Fig. 1, transverse section of limb of leaf (× 14). Figs. 2—4, series of transverse sections through apical region of leaf (× 23) (these sections are from herbarium material, and the exact arrangement of the fused bundles in fig. 4 could not be ascertained). Fig. 5, *Foeniculum vulgare*, Mill. Transverse section of leaf-sheath (× 14). Figs. 6 and 7, *Ficaria verna*, Huds. Fig. 6, transverse section of apex of scale-leaf (× 23). Fig. 7, transverse section of a rather small petiole (× 14). [Arber, A. (1920³).]

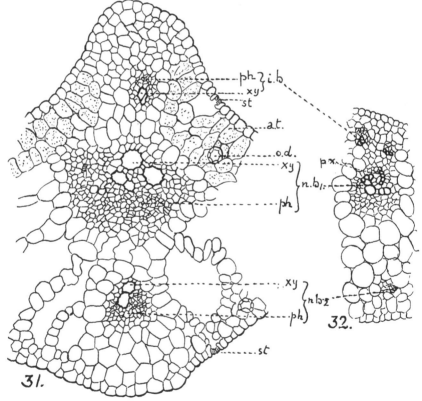

Fig. lvii. *Sagittaria.* (*n.b₁* = bundle of main normal series; *n.b₂* = bundle belonging to second normal series; *i.b.* = inverted bundle; *xy.* = xylem; *ph.* = phloem; *a.t.* = assimilating tissue; *st.* = stomate; *o.d.* = oil duct.) Fig. 31, *Sagittaria sagittifolia*, L. Transverse section of lateral vein of lamina, next but one to midrib (× 400, *circa*). Fig. 32, *Sagittaria montevidensis*, Cham. and Schlecht. Small part of transverse section of leaf near margin in region between arrows in fig. lxxxiii, 11, p. 110 (× 100, *circa*). The lower of the two bundles belonging to the normal series (*n.b₂*) is irregularly placed. Fig. 32 shows that in this genus the inverted bundles are not confined to the ribs. [Arber, A. (1918²).]

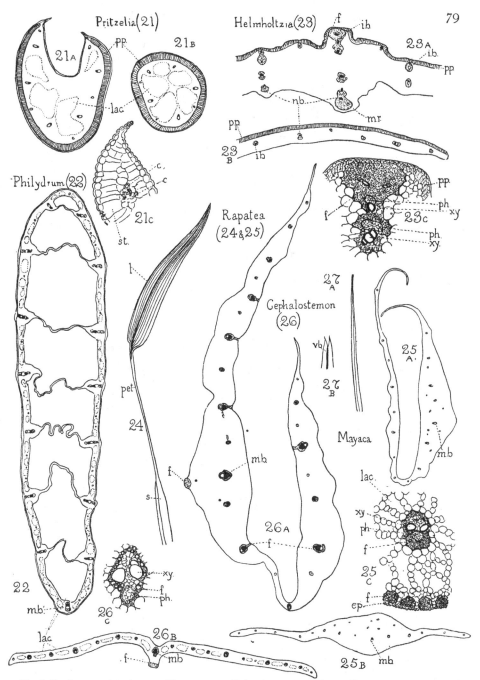

Fig. lviii. Leaves of Farinosae. Fig. 21 A—C, *Pritzelia pygmaea*, F. Muell., transverse section of leaf: A, sheath; B, limb (× 23); C, margin of sheath in A (× 77) to show stomate (*st.*) and crystals (*c.*). Fig. 22, *Philydrum lanuginosum*, Banks, transverse section of limb of leaf (× 14). Fig. 23 A—C, *Helmholtzia acorifolia*, F. Muell., A and B, transverse sections of limb of leaf (× 23): A, midrib region (*m.r.*); B, margin, to show normally orientated bundles and bundle-groups (*n.b.*) and inversely orientated bundles and bundle-groups (*i.b.*); C, inverted bundles of midrib from section similar to A (× 77). Fig. 24, *Rapatea longipes*, Spr., small leaf (× ½). Fig. 25 A—C, *Rapatea angustifolia*, Spr., A and B, transverse section of leaf (× 14); A, sheath; B, limb; C, median bundle (*m.b.*) of limb in B (× 77). Fig. 26 A—C, *Cephalostemon affinis*, Koern.: A, B, transverse section of leaf (× 14); A, sheath; B, limb; C, median bundle of limb (× 77). Fig. 27 A—B, *Mayaca fluviatilis*, Aubl.: A, leaf (× ½); B, leaf apex (× 14); *v.b.*, vascular strand. [Arber, A. (1922[b]).]

completely enclose it; I have observed this process in the leaf of *Triglochin bulbosa*, L., and in the coleoptile of certain Grasses[1]. That there is no hard-

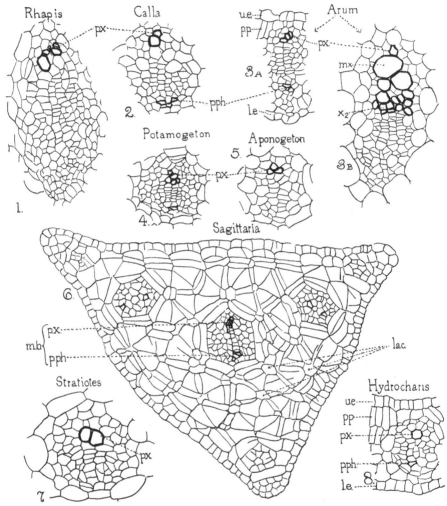

Fig. lix. Intrafascicular Cambium. (Lettering throughout as follows: *p.x.*, protoxylem; *p.ph.*, proto-phloem; *p.p.*, palisade parenchyma; *u.e.*, upper epidermis; *l.e.*, lower epidermis.) Fig. 1, *Rhapis humilis*, Blume, bundle from transverse section of stem close to apex (× 318). Fig. 2, *Calla palustris*, L., bundle from transverse section of petiole of very young leaf (× 318). Fig. 3, *Arum italicum*, Mill.; fig. 3 A, median bundle from transverse section of lamina of very young leaf (× 193); fig. 3 B, lateral bundle from transverse section of petiole of older leaf (× 193); *m.x.*, primary metaxylem; *x₂*, second-ary xylem. Fig. 4, *Potamogeton natans*, L., lateral bundle from transverse section of very young petiole (× 318). Fig. 5, *Aponogeton distachyum*, Thunb., principal bundle from transverse section of very young inflorescence-axis (× 318). Fig. 6. *Sagittaria sagittifolia*, L., transverse section of very young petiole; *m.b.*, median bundle; *lac.*, lacunae, of which only three are yet visible (× 193). Fig. 7, *Stratiotes aloides*, L., bundle from transverse section of base of very young leaf (× 318). Fig. 8, *Hydrocharis Morsus-ranae*, L., median bundle from transverse section of limb of very young leaf (× 318). [Arber, A. (1922³).]

[1] Arber, A. (1924²).

and-fast line between the collateral and concentric types is shown by fig. cviii, 7 A—E, p. 139, in which the history of the single bundle supplying a scale-leaf of *Danae racemosa*, (L.) Mönch., may be followed. The bundle is con-

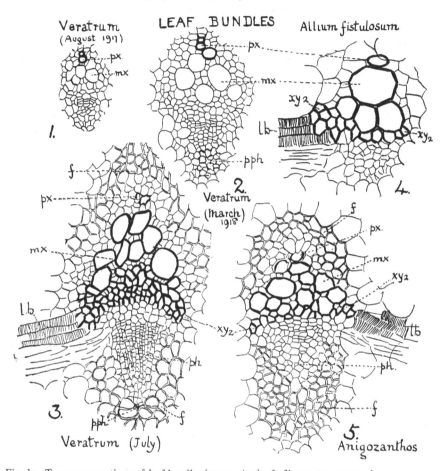

Fig. lx. Transverse sections of leaf-bundles (× 190 *circa*). *f.*, fibres; *px.*, protoxylem; *mx.*, meta-xylem; *xy₂*, secondary xylem; *p.ph.*, protophloem; *ph.*, phloem; *l.b.*, lateral bundle. Figs. 1—3, development of bundle in leaf-blade of *Veratrum album*, L.; fig. 1, from young leaf for next year still enclosed in terminal bud (August 22); fig. 2, from leaf beginning to develop in succeeding spring (March 22); fig. 3, from mature leaf (July 28). Fig. 4, *Allium fistulosum*, L. Fig. 5, *Anigozanthos* sp. Figs. 3—5 show the mode of attachment of the xylem of a branch vein (*l.b.*) to that of the bundle from which it arises. [Arber, A. (1919⁴).]

centric at the base (fig. 7 A), then becomes collateral (fig. 7 B), then con-centric again (figs. 7 C and D), and finally, near the leaf apex, it is again collateral (fig. 7 E)[1].

It is generally assumed in text-books that the leaf-bundles of Mono-cotyledons have no cambium, and hence no secondary xylem or phloem.

---

[1] On concentric bundles in stems, see p. 39, fig. xix, p. 40, and fig. xxi, p. 42.

It would, however, be more correct to say that cambial activity in the leaves of Monocotyledons exists, but is ephemeral. Fig. lix, p. 80 shows very young bundles from the leaves of several Monocotyledons, and it will be recognised that there is meristematic activity between the wood and bast. In fig. lix, 3 B, the cambium can distinctly be seen to have given rise to a few elements of secondary wood. This secondary xylem of Monocotyledonous leaves is characteristically smaller in calibre than the primary xylem, but it plays an important part, for it seems often (if not always) to be the tissue by means of which the xylem of the primary and secondary veins are brought into communication. In fig. lx, 3, p. 81—a section of the

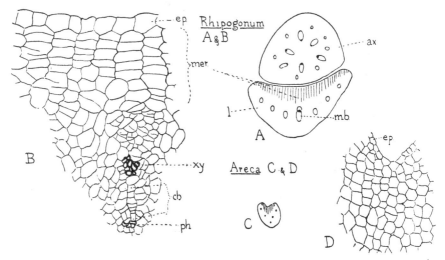

Fig. lxi. Petiolar meristems. A and B, *Rhipogonum album*, R. Br. (Liliaceae); A, transverse section of leaf, *l.*, just above its attachment to axis, *ax.*; *m.b.*, median bundle; *mer.*, subepidermal meristem, the files of cells indicated by parallel lines (× 47); B, upper median region of leaf *l.* in A, to show meristem in greater detail; *xy.*, xylem; *ph.*, phloem; *cb.*, files of cells showing evidence of cambial activity (× 318). C and D, *Areca sapida*, Soland. (Palmae); C, transverse section of petiole (× 14), parallel lines indicating meristematic region, the median part of which is shown in D (× 193). [A.A.]

leaf of *Veratrum*—a transverse lateral vein, *l.b.*, is seen to be attached to the secondary xylem of a longitudinal bundle.

Meristematic activity in Monocotyledonous leaves is not limited to the cambial layer of the bundles. There is often a strong tendency to cell-division towards the adaxial surface of leaf-bases and petioles, and a slight tendency to the same activity towards the abaxial surface. Lonay[1] has shown that in the leaf of *Ornithogalum caudatum*, Ait., not only parenchyma, but also belated vascular bundles are produced by the elements of the leaf-sheath, which "sont douées d'une vitalité remarquable." In the scale-leaf of *Semele androgyna* (fig. cxiii, 36 B, p. 145) the bundles lie so near the

[1] Lonay, H. (1902), pp. 23 and 29; see also Deinega, V. (1898); Bouygues, H. (1902).

ventral surface that the protoxylem region becomes compressed by the multiplication of the adjacent parenchyma cells. As a rule, however, the çell-divisions affect the mesophyll alone. Fig. lxi A shows the leaf-stalk of *Rhipogonum album*, R.Br. (Liliaceae), just above its departure from the axis; the parallel lines indicate the region of meristematic activity. In fig. lxi B a small part of the meristem is drawn on a larger scale. It will be seen that files of five or six elements have been produced by the sub-epidermal layer; in other sections the succeeding cortical layer appeared to be also dividing. The same feature is illustrated for the Palm, *Areca sapida*, Soland., in figs. lxi, C and D, and for *Joinvillea elegans*, Gaudich. (Flagellariaceae), in figs. lxiv, 18 A and B, p. 86. This tendency to cell-multiplication in the leaf-base may be one of the factors that has led to bulb production in some families of Monocotyledons.

A peculiarity of certain Monocotyledonous leaves, for which we can find no exact analogy among Dicotyledons, is the occurrence of tendril leaf-tips Apices of this character often show marked fibrosis. Figs. lxii, 1—3, p. 84, illustrate the tendril structure of the Liliaceous genus *Gloriosa*, in which the bundles are embedded in fibrous tissue. The related genera, *Littonia* (figs. lxii, 4 and 5) and *Sandersonia* (figs. lxiii, 6 and 7, p. 85), have tendrils of a closely similar type, but in the other two Liliaceous tendril-bearers, *Polygonatum* (figs. lxiii, 8 and 9) and *Fritillaria* (figs. lxiii, 10 and 11), the vascular strands, as far as my observations go, remain distinct instead of being embedded in a continuous mass of fibres. *Flagellaria indica*, L. (Flagellaria-ceae, figs. lxiv, 13 and 15, p. 86), has a leaf-tip tendril, which differs from all those found in the Liliaceae in coiling upwards instead of downwards; the relation of the bundles and fibres is also unlike anything met with in Liliaceous tendrils, as will be seen on comparing figs. lxiv, 15 A—C, with the corre-sponding sketches in figs. lxii and lxiii.

Towards the apex of Monocotyledonous leaves there is a tendency to a relative increase in the xylem; in fact the vascular system of the leaf at the extreme apex is generally reduced to wood alone. This apical xylem development is sometimes treated as a special adaptation, related to the extrusion of water, but as it is a conspicuous feature in the genus *Triglochin*[1], in which there are no apical openings or terminal water-pores, I do not think that this view can be maintained.

In the leaf-limb of *Musa Basjoo*, Sieb. (fig. lxv, p. 87), a mass of tracheids, wing-like in section, is developed laterally from the outer side of each of the marginal bundles; as the long cylindrical leaf-tip is reached, these tracheal wings increase in size, approach one another, and finally fuse into one.

But it is not only the tips of Monocotyledonous leaves which are liable to fibrosis—this character is widespread in the foliage of plants of this class, as I

[1] Arber, A. (1924[2]).

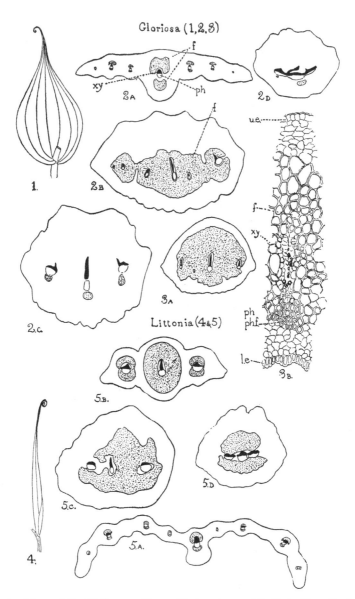

Fig. lxii. Leaf-tip tendrils of Liliaceae. Fig. 1, *Gloriosa virescens*, Lindl., leaf from the Sudan, Kew Herbarium (× ⅔). Figs. 2 and 3, *Gloriosa superba*, L. Figs. 2 A—D, series of sections through leaf-apex and tendril from below upwards (× 21). Figs. 3 A and B, thickened region of another tendril; fig. 3 A, transverse section of tendril (× 10); fig. 3 B, median bundle from a transverse section of the same tendril, at a level where there is slightly less fibrosis than in fig. 3 A (× 141): *ph. f.*, fibres outside phloem. Fig. 4, *Littonia modesta*, Hook., leaf (× ⅔). Figs. 5 A—D, *Littonia Keiti*, Leicht. (*L. modesta*, Hook.), specimen from Temberland, Kew Herbarium, series of transverse sections upwards from below through one leaf apex and tendril (× 21). [(Arber, A. (1923²).]

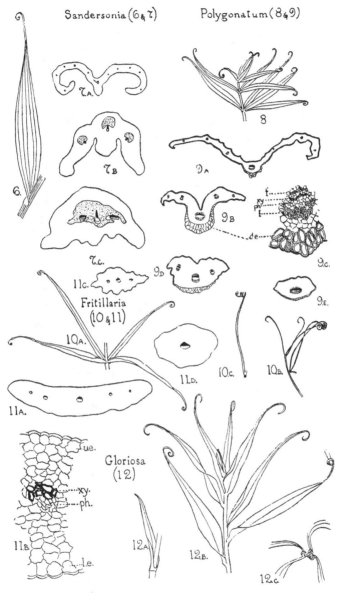

Fig. lxiii. Leaf-tip tendrils of Liliaceae. Figs. 6 and 7, *Sandersonia aurantiaca*, Hook. Fig. 6, leaf of specimen from Swaziland, Kew Herbarium (× ⅔). Fig. 7 A, specimen from Natal, transverse section near leaf apex (× 10); figs. 7 B and C, specimen from Swaziland, transverse sections nearer apex of leaf than fig. 7 A (× 34). Figs. 8 and 9, *Polygonatum sibiricum*, Delar. Fig. 8, apex of shoot with young leaves, Kew Herbarium (× ⅔). Figs. 9 A, B, D, E, series of transverse sections of apical region of leaf, from below upwards, including tendril (× 10); fig. 9 C, median bundle of fig. 9 B on a larger scale (× 34); *d.e.*, thick-walled dorsal epidermis. Figs. 10 and 11, *Fritillaria verticillata*, Ledebour. Fig. 10 A, a node with four leaves, slightly reconstructed, since the herbarium material used was imperfect (× ⅔); figs. 10 B and C, examples of very narrow leaves from the Cambridge Botany School Herbarium (× ⅔). Figs. 11 A and D, transverse sections through one tendril (× 21); fig. 11 B, median bundle of fig. 11 A (× 86); fig. 11 C, transverse section through another tendril (× 10). Fig. 12, *Gloriosa superba*, L. (garden variety). Fig. 12 A, second aerial leaf of shoot (× ⅔); fig. 12 B, shoot apex (× ⅔); fig. 12 C, tips of four leaves with tendrils interlocked (× ⅔). [Arber, A. (1923²).]

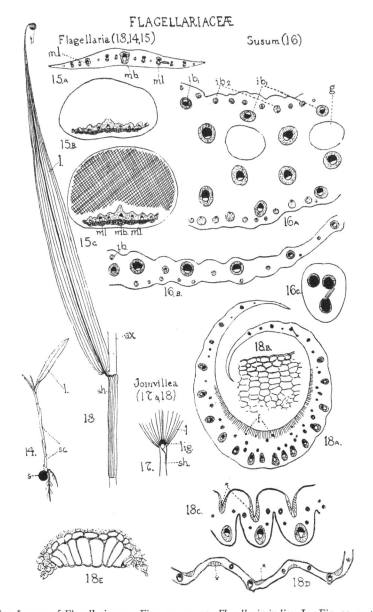

Fig. lxiv. Leaves of Flagellariaceae. Figs. 13, 14, 15, *Flagellaria indica*, L. Fig. 13, mature leaf ($\times \frac{2}{9}$); *ax.*, axis; *sh.*, sheath; *l.*, limb; *t.*, tendril. Fig. 14, seedling with seed (*s.*); scale-leaves (*sc.*); and foliage leaves (*l.*). Fig. 15 A, transverse section of a leaf near apex just below tendril; *m.b.*, median bundle; *m.l.*, main laterals ($\times$ 10); fig. 15 B, transverse section of a somewhat thickened tendril ($\times$ 10); fig. 15 C, transverse section of a tendril which has coiled and thickened; the cross-hatching indicates mesophyll cells which have become fibrous ($\times$ 10). Fig. 16, *Susum anthelminthicum*, Blume. Fig. 16 A, transverse section of a small part of sheath region of leaf ($\times$ 10); *i.b₁* and *i.b₂*, inversely orientated vascular bundles; *g.*, ? gum spaces; fig. 16 B, transverse section of small part near margin of sheath region of leaf ($\times$ 10); fig. 16 C, transverse section near tip of leaf borne on the inflorescence axis of *Susum*? *anthelminthicum* ($\times$ 10). At this level the chief part of the bundle is formed by xylem; at a higher level the three bundles fuse. Figs. 17 and 18, *Joinvillea elegans*, Gaudich. Fig. 17, base of leaf-limb ($\times \frac{2}{3}$); *sh.*, sheath; *lig.*, ligule (indicated in black); *l.*, plicate limb. Fig. 18 A, transverse section of sheath, hairs omitted ($\times$ 6); *f.*, groups of fibres; the lines at right angles to the upper epidermis indicate the radial files of elements, due to meristematic activity; fig. 18 B, part of transverse section towards ventral surface more enlarged to show radial files of cells below upper epidermis ($\times$ 34); fig. 18 C, part of transverse section of lower part of limb, to show origin of plication by invagination, hairs omitted; the upper epidermis has enlarged cells (marked with a cross) at the base of the invaginations ($\times$ 10); fig. 18 D, part of transverse section of limb, not far from apex, hairs omitted; groups of enlarged cells in both upper and lower epidermis; fig. 18 E, group of enlarged cells from lower epidermis, similar to left-hand group marked with a cross in fig. 18 D, more highly magnified ($\times$ 86). [Arber, A. (1923²).]

have learned to my cost in cutting sections. The Xyridaceae (fig. lxvi, p. 88) furnish an extreme instance. The epidermis in the genus *Xyris* is usually much sclerised, and, in some cases, the elongation of these thickened cells results in the formation of a conspicuous border at the leaf-margins (e.g. *X. anceps*, figs. lxvi, 14 A—D). Masses of fibres are also associated with the bundles.

The leaves borne by corms and rhizomes in various Monocotyledonous families sometimes possess nerves of so fibrous a character that on the decay of the softer parts they are left as a relatively permanent fringe or basket-work, ensheathing the axis, e.g. *Cyanella* sp., fig. xxiii, 2, p. 46. In fig. 2 C, the leaf enclosing the corm, c_2, will, presumably, in the course of the following season, be reduced to a fibrous cup, such as that which envelops the corm, c_1, in figs. 2 A and B.

One is tempted to wonder whether the excessive sclerosis of Monocotyledonous leaves is to be correlated with the relatively poor development of their

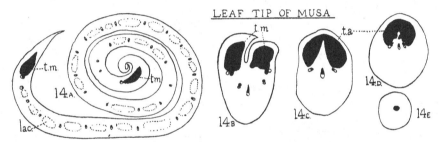

Fig. lxv. *Musa Basjoo*, Sieb. Series of transverse sections through tip of one foliage leaf (× 23). 14 A, section of limb a little below solid apex; the limb is rolled in the bud in a counter-clockwise direction, but as this series is taken upwards from below, the direction appears reversed; *t.m.*, tracheal mass; *lac.*, lacuna; 14 B, section at transition to solid apex; 14 C and D show fusion of tracheal masses to an arc, *t.a.* In 14 E the vascular system is reduced to a small group of elements, which seem to be chiefly tracheids. [Arber, A. (1922[1]).]

xylem, which is due to the minimal character of their secondary thickening. However this may be, we find, when we consider the individual bundles, that stages can be traced between normal collateral strands, with phloem outside the xylem, and strands in which the vascular elements have altogether disappeared, and the fibres alone remain[1]. I noticed, for instance, on comparing the leaf-sheath of a plant of *Triglochin maritima* from England with one from the Andes, that certain strands which, in the English example, were provided with xylem and phloem, were represented in the plant from South America by fibres alone[2]. It thus seems possible that some of the non-vascular fibrous strands, which occur regularly in the leaves of certain Monocotyledons, may be, perhaps, degraded vascular bundles.

It is usual in systematic botany to treat the vegetative organs as of comparatively little value as guides to affinity, and thus to concentrate

[1] Drabble, E. (1906); Arber, A. (1918[2]), p. 493.
[2] Arber, A. (1924[2]), figs. 4 A and B; see also figs. 5 D—G.

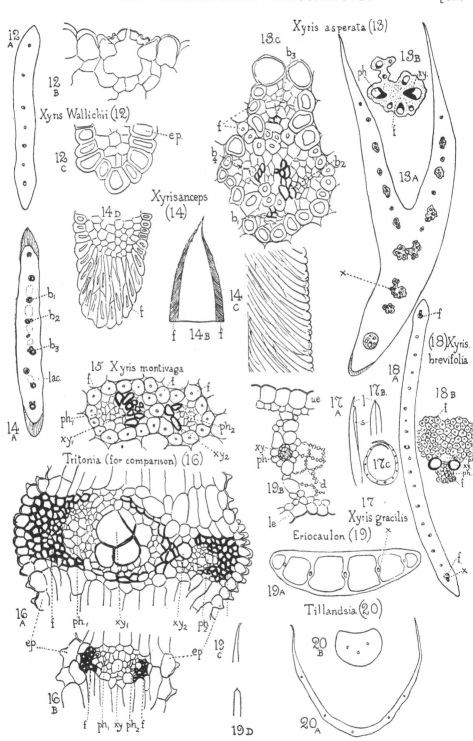

Xyris asperata (13)

Xyris Wallichii (12)

Xyrisanceps (14)

15 Xyris montivaga

Tritonia (for comparison) (16)

(18)Xyris brevifolia

Xyris gracilis (17)

Eriocaulon (19)

Tillandsia (20)

Fig. lxvi. Leaves of Farinosae. Figs. 12 A—C, *Xyris Wallichii*, Kth.: A, trans. sect. leaf-limb (× 23); B, stomate from A (× 193); C, margin of A (× 193). Figs. 13 A—C, *Xyris asperata*, Kth. (*trachyphylla*, Mart.): A, trans. sect. transition region between sheath and limb (× 23) to show bundle-groups; B, bundle-group marked × in A (× 77); C, group of four bundles (b_1, b_2, b_3, b_4), in common fibrous sheath, from transverse section of leaf-limb (× 318). Figs. 14 A—D, *Xyris anceps*, Lam.: A, transverse section limb of leaf (× 23); b_1, single bundle; b_2, group of two bundles; b_3, group of three bundles; B, apex of leaf-limb viewed as solid object (× 23); C, fibrous margin of leaf viewed as solid object (× 77); D, transverse section margin of A (× 77). Fig. 15, *Xyris montivaga*, Kth.: pair of opposite bundles (xy_1 and ph_1, xy_2 and ph_2) in common fibrous sheath (× 318); similar to b_2 in fig. 14 A (*X. anceps*). Figs. 16 A, B, *Tritonia* (garden hybrid) for comparison, lettering as in fig. 15: fibres represented in black (instead of white with double lines indicating thickness of walls, as in other figures): A, pair of opposite bundles from transverse section of leaf-limb (× 318); B, similar pair of bundles, but with xylems fused (× 318). Figs. 17 A—C, *Xyris gracilis*, R.Br.: A, leaf consisting mainly of leaf-sheath, with reduced limb (*l.*); B, apex of another reduced leaf in which limb is minute (both natural size); C, transverse section of leaf shown in B (slightly enlarged). Figs. 18 A—B. *Xyris brevifolia*, Mich.: A, transverse section leaf-limb (× 23); B, marginal bundle such as that marked × in A (× 193). Figs. 19 A—D, *Eriocaulon*: A and B, *E. septangulare*, With.; A, transverse section of limb of leaf (× 23); B, lamella marked × in A (× 77); *u.e.*, upper epidermis, *l.e.*, lower epidermis; *d.*, fragment of diaphragm seen in surface view; C, *E. Wallichianum*, Mart., f. *submersa*, apex of leaf (× ½); D, *E. cuspidatum*, Dalz., apex of leaf to show mucro (× ½). Figs. 20 A, B, *Tillandsia usneoides*, L., transverse section leaf (× 23): A, sheath; B, limb. [Arber, A. (1922^b).]

attention on the flowers and fruits. But those who have made a detailed study of the comparative anatomy of Monocotyledons seem to be unanimous in concluding that, in many cases, genera and even species can be determined on structural grounds without any help from the reproductive organs. It might, perhaps, have been supposed that leaf-anatomy, for instance, would vary so much with the conditions of life, that ancestral characters would be liable to be obliterated. But in practice this is not found to be the case. The great majority of the marine Monocotyledons, for example, though all developing in a similar and very special environment, can be identified specifically from sections of the leaves alone[1]. And—at the other extreme of the scale of habitats—Curtis's studies of the leaves of the epiphytic Orchids of New Zealand[2] show how striking may be the anatomical variations among a group of plants belonging to a single family, and living under corresponding conditions. Again, in the genus *Agave*, a number of species which seldom or never bloom are cultivated in Central Europe, and a botanist, who examined them to see how far identification could be carried without the help of the flowers, found that the very numerous species with which he was concerned could be recognised by leaf-characters alone[3].

As an example of specific differences in anatomy, the leaf-structure of several species belonging to the Section *Juno* of the genus *Iris* is illustrated in figs. lxvii, 1—5, p. 90. The characters, by which these species are found to be distinguishable, are the absence or presence of epidermal papillae, and their distribution, if present, and also the form of the fibrous girders associated with the vascular bundles. It must not be assumed, however, that species can *always* be discriminated by their leaf-anatomy. I have failed, for

---

[1] For references see Arber, A. (1920^4), p. 131.　　　[2] Curtis, K. M. (1917).

[3] Müller, C. (1909).

instance, to find any structural character in the leaf which could be used to differentiate two of these *Juno* Irises—*I. bucharica,* Foster (fig. lxvii, 5), and

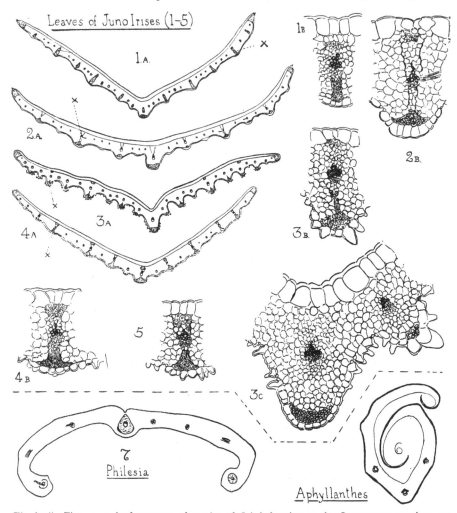

Fig. lxvii. Figs. 1—5, leaf-structure of species of *Iris* belonging to the *Juno* group; sections cut from young leaves gathered in February. Figs. 1 A and B, *I. persica*, L., var. *stenophylla*; fig. 1 A, transverse section near apex (× 14); fig. 1 B, vein marked with a × in fig. 1 A (× 77). Figs. 2 A and B, *I. Rosenbachiana*, Regel; fig. 2 A, transverse section (× 14); fig. 2 B, vein marked with a × in fig. 2 A (× 77). Figs. 3 A—C, *I. sindjarensis*, Boiss.; fig. 3 A, transverse section near apex (× 14); fig. 3 B, vein marked with a × in fig. 3 A (× 77); fig. 3 C, midrib and next vein in fig. 3 A (× 77). Figs. 4 A and B, *I. orchioides*, Carrière; fig. 4 A, transverse section near apex (× 14); fig. 4 B, bundle marked with a × in fig. 4 A (× 77). Fig. 5, *I. bucharica*, Foster, transverse section of vein corresponding to that of *I. orchioides* shown in fig. 4 A (× 77). Fig. 6, *Aphyllanthes monspeliensis*, L., transverse section of leaf (× 23). Fig. 7, *Philesia magellanica*, J. F. Gmel., transverse section of leaf (× 14). [A.A.]

I. orchioides, Carrière (figs. lxvii, 4 A and B)—though, possibly, a more thorough examination might reveal some distinction.

CHAPTER V

THE FOLIAGE-LEAF: INTERPRETATION

"Les feuilles des Monocotylés montrent une assez grande diversité;...leur
étude promet au morphologiste de curieux résultats." D. CLOS, 1875.

(i) *The Leaf-skin Theory*

THE term "leaf" is one for which it is by no means easy to find an adequate
definition, but both botanists and laymen have, until recently, generally
agreed in regarding this organ as an appendage which—except for a cir-
cumscribed area of attachment—is completely distinct from the axis which
bears it. Within the last few years, however, a theory has been put forward
by E. R. Saunders[1], which gives a different picture of the shoot in the
Flowering Plants. According to this hypothesis—the "Leaf-skin Theory"—
the axis is entirely clothed by a mosaic-like covering, each element of which
consists of the downward prolongation of a single leaf-base. Something of the
kind had been previously recognised for those plants whose leaf-bases are
"decurrent," but Saunders has shown that, even where the leaf-bases
have no obviously decurrent character, the fact that the "stem" surface
is made up of units each belonging to an individual leaf can be demonstrated by the distribution of hairs, and by other features of the surface
topography. When one's mind is once opened to this idea, the "leaf-skin"
nature of the stem-superficies is perhaps even more easily recognisable in
Monocotyledons than in Dicotyledons. The axes of *Sisyrinchium*, drawn
in the accompanying figure (figs. lxviii, 25 A—E, p. 92), furnish a case of rather
an exaggerated nature, in which downward prolongations of the leaves
form conspicuous wings. Examples of a more typical kind may often be
seen on examining the flowering shoots of the Crown-imperial, *Fritillaria
Imperialis*, L. In the shoot drawn in fig. lxix, p. 93, for instance, the region of
the inflorescence-axis above the foliage leaves was a dull purplish colour
(black in the sketch) and thus contrasted sharply with the lower region,
which was completely clothed with green decurrent leaf-bases, whose identity
was perfectly distinct on account of their slightly winged margins. It was
noticeable that, in the region *b*, only the right-hand side of the axis had a
leaf-base coat of a green hue—a bizarre effect being thus produced. I came
upon a still more curious instance in another plant examined at the same
time; here one solitary leaf was located above all the others, and, from its

[1] Saunders, E. R. (1922).

Winged Axes

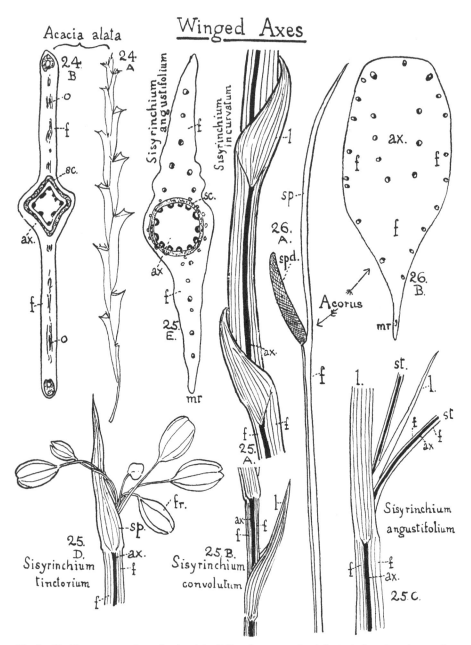

Fig. lxviii. Figs. 24 A and B, *Acacia alata*, R.Br.; fig. 24 A, shoot (nat. size) to show bases of 2-ranked phyllodes decurrent through two internodes, producing winged axis; fig. 24 B, transverse section of winged axis (× 28); *ax.*, axial region; *f.*, foliar region; *sc.*, fibrous hollow cylinder; *o.*, bundles cut obliquely. Figs. 25 A—E, *Sisyrinchium*; *ax.*, axis (black), and *f.*, foliar wing of stem; fig. 25 A, *S. incurvatum*, Gardn. (nat. size), axis with equitant leaves (*l.*) reduced almost to sheathing bases; fig. 25 B, *S. convolutum*, Nocca (nat. size); fig. 25 C, *S. angustifolium*, Mill., part of branched axis (nat. size); fig. 25 D, *S. tinctorium*, H. B. et K., top of infructescence-axis (nat. size) with spathe, *sp.*, and fruits, *fr.*; fig. 25 E, *S. angustifolium*, Mill., transverse section of winged axis (× 28); *m.r.*, continuation of dorsal margin of leaf next above; *sc.*, fibrous hollow cylinder. Figs. 26 A and B, *Acorus Calamus*, L.; fig. 26 A, fertile axis, much reduced; *spd.*, spadix; *sp.*, spathe; *f.*, winged side of axis; fig. 26 B, transverse section of winged axis below spathe (× 7); *m.r.*, continuation of midrib region of spathe. It should be noted that the parts in these diagrams labelled as "axial" are, in reality, clothed with a leaf-skin. [Arber, A. (1921[1]).]

base, a single strip of green "leaf-skin" ran down the brown axis. It must
be realised, however, that this smooth dark-coloured part of the axis, though
to all appearance "naked," is itself, in reality, clothed with a leaf-skin, prolonged downwards from the bases of the cluster of leaves associated with
the flowers; these leaves are shown in fig. cliv, I A, p. 202.

If we accept the leaf-skin theory, we cannot escape the conclusion that
the leaf plays a much more important rôle in the organisation of the shoot

Fig. lxix. *Fritillaria Imperialis*, L., segment of flowering shoot, April 28, 1919 (× ½), to show contrast between, *a.*, region of inflorescence-axis above foliage leaves, in which the epidermis is a dull
purplish colour; *c.*, lower region of axis, completely clothed with green, decurrent leaf-bases;
and *b.*, transition region. [A.A.]

than has hitherto been generally held. The leaf, as the text-books understand it, may be compared with the *visible* parts of an iceberg, whereas the
leaf, considered in the light of the leaf-skin theory, corresponds not only to
the visible iceberg, but, in addition, to that part which does not emerge
above the sea-level, and which, we are told, greatly exceeds the part which
rises into the air. The leaf-skin conception alters many of our ideas; it
gives us, for instance, a different picture of those plants which are commonly described as almost "leafless." The rhizomes of *Corallorhiza* and

Epipogum (Orchidaceae), for example, bear scales which generally consist of three cell-layers only, and are scarcely visible[1]. But we have now to reckon, as part of the leaf-system, not only these insignificant scales, but also the entire leaf-skin surface of the rhizome. It is noticeable, again, that even in such cases as *Danae racemosa*, (L.) Mönch., the Victor's-laurel, and

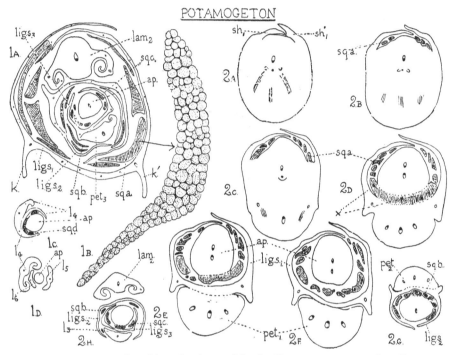

Fig. lxx. *Potamogeton* sp. (broad-leaved, submerged form). Fig. I A, transverse section of inner part of apical bud (×47); in the outermost leaf represented, only the ligular sheath (*lig.* s_1) with its keels (*k.* and *k'.*) are shown; limb (*lam.* 2) and ligular sheath (*lig.* s_2) of second leaf; petiole or base of limb (*pet.* 3) and sheath (*lig.* s_3) of third leaf; *ap.*, shoot apex. Three sets of squamules (shaded), *sq. a*, *sq. b*, *sq. c*, are seen outside second leaf, third leaf, and apex respectively. Fig. I B, transverse section of squamule marked with arrow in fig. I A (×193). Figs. I C and D, changes in shoot apex (*ap.* of fig. I A) as tip is approached (×47). Fig. I C, fourth leaf (l_4) detaching itself from apex, and squamules (*sq. d*) occurring between it and apex. Fig. I D is close to tip of shoot, and shows highest leaves (l_5 and l_6) on either side of apex (*ap.*). Figs. 2 A—H, series of transverse sections from below upwards through three successive young leaves and growing-point of another bud (×47); sections broken on side towards petiole of leaf 1, so this region reconstructed; lettering as in fig. I; sh_1 and sh'_1, margins of sheathing leaf-base. [Arber, A. (1923[1]).]

Ruscus aculeatus, L., the Butcher's-broom, in which the leaves are reduced to ephemeral scales, their bases can yet be traced down the internodes as conspicuous ridges (figs. xvii, I A and 2 A, p. 38, and fig. cviii, 6 A, p. 139). The leaf-skin theory thus prevents our treating the foliar part of the shoot as negligible, even in plants in which the free part of the leaf is reduced to its lowest terms. There are, however, many Monocotyledons in which, though

[1] Velenovský, J. (1907), p. 548.

POTAMOGETON

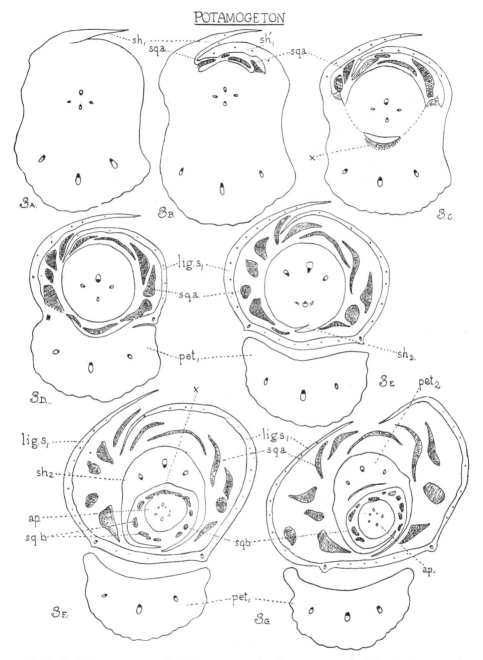

Fig. lxxi. *Potamogeton natans*, L. Figs. 3 A—E, series of transverse sections from below upwards through apical bud including two successive leaves and the squamules within them (× 47); lettering as in fig. lxx on opposite page. Figs. 3 F and G show further stages in the development of the second leaf and the squamules. Fig. 3 F, from a broken section, somewhat reconstructed as regards *lig. s₁* and *sq. a* on right. [Arber, A. (1923[1]).]

the leaves—using the term in its ordinary sense—are well developed, the internodes are much abbreviated, and hence the area of leaf-skin corresponding to each leaf is reduced almost to vanishing point; these cases may be treated as the converse of those in which the leaf-skin clothes a relatively long internode, and in which the free part of the leaf is insignificant when compared with its downward prolongation (fig. lxviii, 25 A, p. 92).

In connexion with the leaf-skin theory, a certain interest attaches to the curious little scales, known as "squamulae intravaginales[1]," associated with the leaf-bases of the Helobieae and some Araceae, which are illustrated in figs. lxx—lxxiii, pp. 94–97. These scales have often been regarded as axillary to the leaves, but it now seems probable that they do not belong to the leaf in whose axil they sometimes seem to arise, but that they are, in reality,

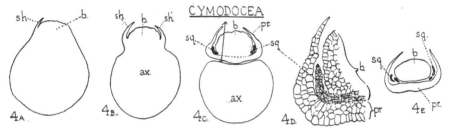

Fig. lxxii. *Cymodocea isoëtifolia*, Asch. Figs. 4 A, B, C, E, series of transverse sections (× 23) passing upwards from below through an axillary bud, *b*. (whose farther history has not been followed), and its prophyll, *pr.*; the bud is lateral to an axis, *ax*. Figs. 4 A and B show detachment of margins of prophyll, *sh*. and *sh′*. Fig. 4 C, complete detachment of lateral bud, from which the prophyll is at this level only partly free; dotted line indicates plane of separation between prophyll and axis of bud; *sq.*, squamule. Fig. 4 D, region to left of arrow in fig. 4 C (× 77). Fig. 4 E, stage at which bud, squamules, prophyll, and parent axis have become free from one another. [Arber, A. (1923[1]).]

appendages of the basal region of the leaf-skin belonging to the next leaf above. This is particularly clear in *Triglochin*. In figs. lxxiii, 5 A—C, the origin of some of the scales can be followed; it will be noticed, for instance, that those inside the leaf l_1 are not axillary to it, but are budded off from the surface of the succeeding internode—in other words, from the leaf-skin which forms the downward prolongation of the next leaf, l_2.

(ii) *Stipules and Ligules*

The problem of the relation to one another of the different types of leaf-base met with among Monocotyledons—ranging from the simple sheath to elaborately stipulate or ligulate forms—is still a subject of controversy. The interpretation, which seems to me the most reasonable, is indicated in the accompanying diagrams (fig. lxxiv, p. 98), in which the petiole or limb is

---

[1] Arber, A. (1923[1]), and references in this paper to earlier work; see also Arber, A. (1925).

indicated in black, and the leaf-sheath in white, while the outgrowths forming the stipules and ligules are dotted. I regard the simple leaf-sheath (A)
as the fundamental type, from which stipules or ligules are developed by a
secondary process of winging. The variations in leaf-base form met with
among Monocotyledons depend upon the proportions borne by the sheath
to its wing-like outgrowths, and upon the position of the latter, and their
freedom or fusion; the part that may be played by these factors is shown
in fig. lxxiv. By a simple upward extension of the leaf-sheath, where it
abuts on the petiole, a ligulate type (B) is easily derivable from A. The

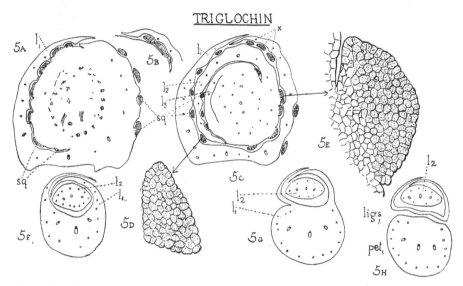

Fig. lxxiii. *Triglochin maritima*, L. Figs. 5 A, B, C, F, G, H, serial transverse sections from below
upwards through young leaf, l_1, and the apical bud with younger leaves, l_2 and l_3, which it encloses; squamules, *sq.*, shaded ($\times$ 23). Fig. 5 A, margin of leaf-sheath of l_1 free on left side, but,
owing to slight obliquity of section, cut at a lower level on right-hand side, and there fused with
axis. Fig. 5 C, section at a slightly higher level, showing three sets of squamules, external respectively to l_1, l_2, and l_3. Fig. 5 B shows the three squamules marked with a cross in fig. 5 C, cut
at a slightly lower level. Fig. 5 D, detached squamule marked with arrow to left of fig. 5 C ($\times$193).
Fig. 5 E, attached squamule marked with arrow to right of fig. 5 C ($\times$ 193). Figs. 5 F, G, H,
sections through l_1 and l_2 at higher levels, to show development up to point of separation of ligular
sheath and petiole of l_1 ($\times$ 23). [Arber, A. (1923[1]).]

upper lateral margins of the leaf-sheath may, on the other hand, grow out
into two wings (C), which, by the reduction of the basal part of the sheath,
may come to form stipules of the free lateral type (D). As a further modification, the wings of the leaf-sheath may fuse with greater or less completeness in front of the petiole base (E and F), and from such types an "axillary"
stipule or ligule (G and H) may be developed by reduction of the rest of
the leaf-base. H may, alternatively, be derived by the same process from
such a form as B. In these diagrams—for the sake of simplicity—only the

open forms of sheath and ligule are represented; the tubular ligule may, however, be visualised as having developed, on corresponding lines, through the upward winging of a tubular leaf-sheath.

A view totally different from that just outlined has been maintained by Glück[1], who treats the closed leaf-sheaths, so numerous among Monocotyledons, as *stipulae adnatae*, which have arisen by the fusion of a pair of free stipules with the base of the petiole. This interpretation, which deprives the leaf-sheath of its rank as a primary element of the leaf, seems to me

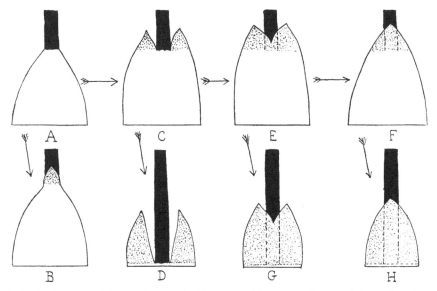

Fig. lxxiv. Diagram of different forms of leaf-base met with amongst Monocotyledons, viewed from adaxial side, to illustrate development of stipules and ligules, from the simple form of leaf-sheath, A. Petiole or limb indicated in black; leaf-sheath, white; stipular or ligular wings dotted. In the region where the ligule or stipular wings are adaxial and free from the petiole, the outline of the latter is indicated with a broken line, as if the ligule were semi-transparent. See pp. 96, 97. [A.A.]

highly artificial; it is hardly possible, I think, to suppose that every one of the countless sheathing leaf-bases met with in this class has had a stipular origin. In the Iridaceae, for instance, there are no examples of stipules, and only one genus, *Geissorhiza*, is known to possess a ligule. It is difficult to imagine that all the simple leaf-sheaths of the other fifty-six genera, belonging to the family, have originated from stipules of whose existence no trace whatever remains.

---

[1] Glück, H. (1901), p. 81; on stipules see also Colomb, G. (1887).

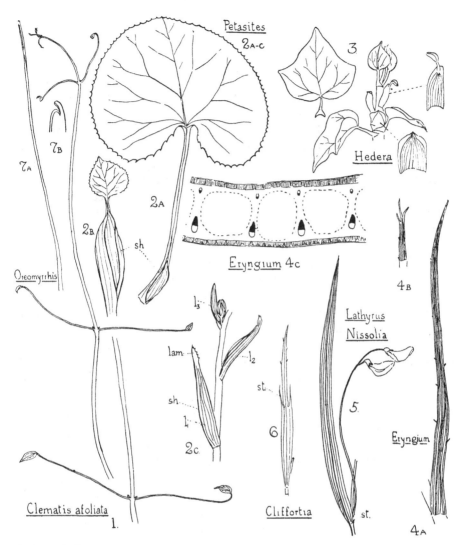

Fig. lxxv. Reduced leaves among Dicotyledons. Fig. 1, *Clematis afoliata*, Buch., shoot (cut in two) showing laminae reduced in comparison with petioles ($\times \frac{1}{2}$). Figs. 2 A—C, *Petasites vulgaris*, Desf., shoot in December ; fig. 2 A, small normal leaf growing from subterranean axis, to show comparatively small part played by sheath ($\times \frac{1}{2}$); figs. 2 B and C, reduced leaves associated with inflorescence ($\times \frac{1}{2}$); fig. 2 B, leaf from near base of inflorescence-axis, with reduced lamina and petiole, but conspicuous sheath ; fig. 2 C, part of inflorescence-axis showing three leaves; l_1 and l_2 have reduced laminae, while in l_3, which has a capitulum in its axil, the lamina is not distinguishable with certainty; in these three leaves, petioles seem to be entirely absent. Fig. 3, *Hedera Helix*, L., terminal bud ($\times \frac{1}{2}$) to show gradation from sheathing scale leaves, in which there is a rudimentary tip, representing a reduced petiole and lamina, to the normal leaf in which the leaf-base is very small. Figs. 4 A—C, *Eryngium* ; figs. 4 A and B, *E. ebracteatum*, Lam.; fig. 4 A, foliage leaf; fig. 4 B, small leaf associated with inflorescence ($\times \frac{1}{2}$); fig. 4 C, *E. pandanifolium*, Cham. et Schlect., transverse section of part of leaf ($\times 14$) to show inverted bundles towards upper surface. Fig. 5, *Lathyrus Nissolia*, L., linear, parallel-veined leaf with minute stipules, *st.*, at base. Fig. 6, *Cliffortia graminea*, Linn. f., leaf ($\times 1$) showing sheathing leaf-base with stipules, *st.*, and reduced blade. Figs. 7 A and B, *Oreomyrrhis linearis*, Hemsley, specimen from British New Guinea, Kew Herbarium ; fig. 7 A, small leaf ($\times \frac{1}{2}$); fig. 7 B, apex of leaf enlarged. [A.A.]

(iii) *The Phyllode Theory*

The main peculiarity of the leaves of Monocotyledons, as a class, is, as we have noted already, the recurrence, in family after family, of a type of leaf in which a basal leaf-sheath passes distally into a simple limb traversed by parallel veins. Among Dicotyledons, the characteristic leaf-form is a more complex one, but a close resemblance to the Monocotyledonous type can be traced in a number of "reduced" leaves, in which the lamina is either altogether absent, or represented by the midrib region alone. Fig. lxxv, p. 99, illustrates some of these cases. The parallel-veined foliage of certain species of *Eryngium* (fig. lxxv, 4 A and B) appears to represent the leaf-base and rachis of a compound pinnate leaf, the pinnules surviving, in a rudimentary form, as marginal teeth. Such leaves have a Monocotyledonous look, and may also in some cases recall, in their midrib-like anatomy[1] (fig. lxxv, 4 C), the structure of certain Monocotyledonous leaves with "inverted" bundles towards the adaxial surface, which we shall discuss shortly. But it is important to remember that, in the Monocotyledonous leaf, there are neither reduced pinnules, nor any other sign of a derivation from a compound pinnate form; therefore seems probable that such leaves as those of the *Eryngia* do not furnish a perfectly exact comparison for those of Monocotyledons. The hypothesis that explains the leaves of Monocotyledons by reference to the *Eryngium* type, thus appears to introduce unnecessary complexities; a simpler and more satisfactory Dicotyledonous analogy is offered by certain leaf-sheath leaves and by certain petiolar phyllodes, which—as I hope to show in the following pages—can be precisely paralleled among Monocotyledons, both as regards their form, their venation, and their internal structure. As examples of Dicotyledonous leaf-sheath leaves, I have illustrated the inflorescence-leaves of *Petasites vulgaris*, Desf. (fig. lxxv, 2 C), the bud-scales of *Hedera Helix*, L. (fig. 3) and the foliage leaf of *Oreomyrrhis linearis*, Hemsley (fig. 7); in the last-named, the leaf consists, almost entirely, of an elongated leaf-base, the other regions being represented merely by a minute hooded tip (fig. 7 B)[2]. Reduced Dicotyledonous leaves in which, conversely, the leaf-stalk plays the chief part, are illustrated in *Clematis afoliata*, Buch. (fig. lxxv, 1) and, more strikingly, in the petiolar phyllodes of various Acacias (figs. lxxvi, p. 101, lxxix, p. 105, lxxxv, p. 113, etc.) and those of *Oxalis bupleurifolia*, A. St. Hil. (fig. lxxxv, 3 A and B). It was A. P. de Candolle[3], nearly a hundred years ago, who first saw the significance of the most reduced of these leaves, and suggested, in his classic *Organographie*, that the linear leaf of the Monocotyledon is equivalent to the leaf-base and petiole of the Dicotyledon—the blade being unrepresented. This view—which we may, for brevity, call the "phyllode

---

[1] Gaisberg, E. von (1922).   [2] Domin, K. (1909).   [3] Candolle, A. P. de (1827).

ALLIUM (sections-PORRUM, SCHOENOPRASUM, RHIZIRIDIUM, MACROSPATHA)

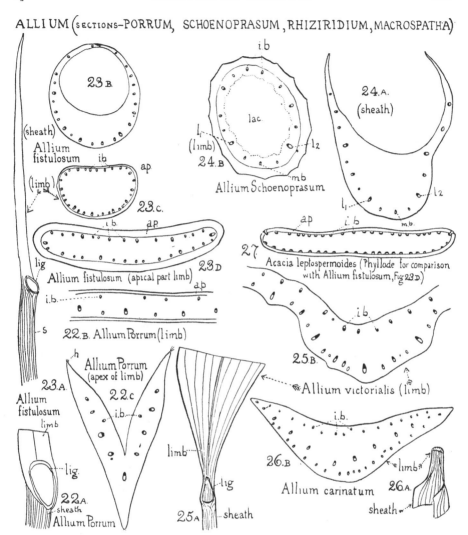

Fig. lxxvi. Leaf structure of *Allium* (xylem, black; phloem, white). Figs. 22 A—C, *Allium Porrum*, L. (Sect. *Porrum*). Fig. 22 A, junction of sheath and limb, with ligule, *lig.* (½ nat. size); fig. 22 B, transverse section of part of limb, not including midrib (× 9, *circa*); *i.b.*, inverted bundle; *a.p.*, assimilating parenchyma; fig. 22 C, transverse section close to apex of limb, to show survival of inverted bundles, *i.b.*, in this region; *h.*, marginal hairs (× 23). Figs. 23 A—D, *Allium fistulosum*, L. (Sect. *Schoenoprasum*). Fig. 23 A, leaf (½ nat. size) to show upper part of sheath, *s.*, ligule, *lig.*, and limb; fig. 23 B, transverse section of sheath (× 5½, *circa*); fig. 23 C, transverse section of limb (× 5½, *circa*); *i.b.*, inverted bundle; fig. 23 D, transverse section of flattened apical part of limb (× 14). Figs. 24 A and B, *Allium Schoenoprasum*, L. Fig. 24 A, transverse section of sheath (× 23); fig. 24 B, transverse section of limb (× 23). (Note in both cases relative unimportance of midrib, *m.b.*, as compared with main laterals, *l*₁ and *l*₂.) Figs. 25 A and B, *Allium victorialis*, L. (Sect. *Rhiziridium*). Fig. 25 A, junction of limb and sheath showing ligule, *lig.* (½ nat. size); fig. 25 B, transverse section of midrib region of limb; *i.b.*, inverted bundle (× 8½, *circa*). Figs. 26 A and B, *Allium carinatum*, L. (Sect. *Macrospatha*). Fig. 26 A, junction of sheath and limb (½ nat. size); fig. 26 B, transverse section of limb near its junction with sheath (× 8½, *circa*). Fig. 27, *Acacia leptospermoides*, Benth., transverse section of phyllode (× 14) for comparison with *Allium fistulosum* (fig. 23 D); *a.p.*, assimilating parenchyma; *i.b.*, inverted bundles. [Arber, A. (1920³).]

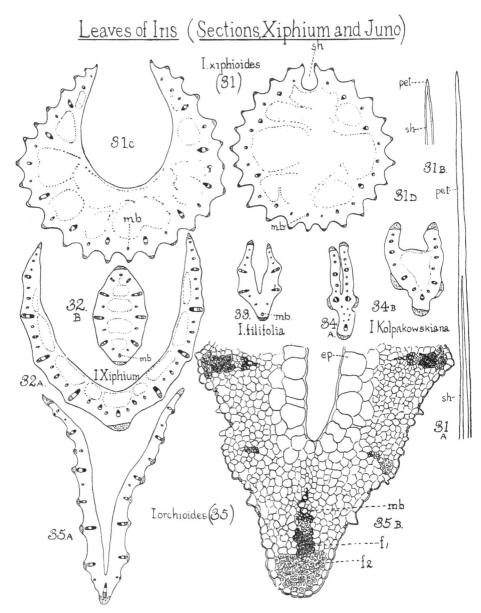

Fig. lxxvii. Figs. 31 A—D, *Iris xiphioides*, Ehr.; figs. 31 A and B, apical region of two leaves seen from ventral side (× ½) to show variation in length of cylindrical apex (*pet.*) which terminates the bifacial region (*sh.*); fig. 31 C, transverse section of bifacial region (× 14); fig. 31 D, transverse section of limb at uppermost limit of junction of sheath (*sh.*) and limb (× 14). Figs. 32 A and B, *I. Xiphium*, L.; fig. 32 A, transverse section of sheath; fig. 32 B, transverse section of apical limb (× 14). Fig. 33, *I. filifolia*, Boiss. (× 11, *circa*). Figs. 34 A and B, *I. Kolpakowskiana*, Regel; fig. 34 A, transverse section of leaf (× 23); fig. 34 B, transverse section nearer apex (× 24, *circa*); the epidermis is thickened and papillose, but this is not shown in these diagrams. Figs. 35 A and B, *I. orchioides*, Carr.; fig. 35 A, transverse section of leaf (× 14); fig. 35 B, midrib region of fig. 35 A (× 77, *circa*). Note contrast between lignified fibres, *f₁*, and non-lignified hypodermal fibres, *f₂*. [Arber, A. (1921[1]).]

theory"—obtained some support in the early years of the nineteenth century, but interest in it, as in so many morphological questions, fell into abeyance

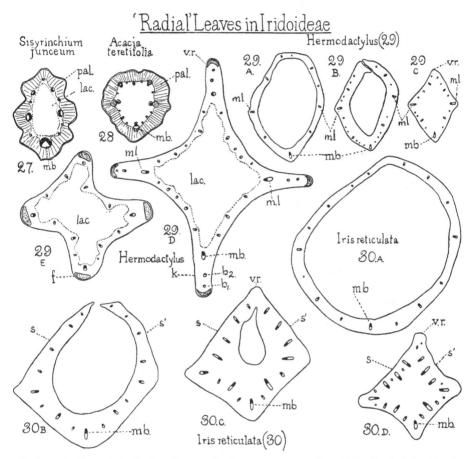

Fig. lxxviii. Fig. 27, *Sisyrinchium junceum*, E. Mey., transverse section of limb of leaf; *pal.*, 3-tiered palisade parenchyma (× 23). Fig. 28, *Acacia teretifolia*, Benth., transverse section of phyllode; *pal.*, palisade parenchyma of about two tiers of cells (× 23). Figs. 29 A—E, *Hermodactylus tuberosus*, Mill.; *m.l.*, main lateral; figs. 29 A—C, sections from microtome series through base of leaf (× 14) showing transition from closed sheath through open sheath to solid limb; *v.r.*, ventral ridge; fig. 29 D, mature part of another leaf; *k.*, keel; b_1 and b_2, bundles of keel; *lac.*, lacuna; fig. 29 E, section close to apex of same leaf; *f.*, fibres; *lac.*, lacuna crossed by trabeculae. Fig. 30 A—D, *Iris reticulata*, Bieb., sections from a series through base of limb of leaf (× 14). Figs. 30 A and B, sheath; fig. 30 C, transition region; fig. 30 D, limb showing asymmetry of the two lateral faces, *s.* and *s'.*, which meet at the ventral ridge, *v.r.* [Arber, A. (1921[1]).]

during the period when Darwin's doctrine of adaptation held paramount sway, and it is only recently that it has again received active consideration[1].

---

[1] Buscalioni, L. and Muscatello, G. (1908); Henslow, G. (1911); Arber, A. (1918[2]) and later papers; Menz, G. (1922); Ghose, S. L. (1923). For criticism see Glück, H. (1919); Bugnon, P. (1921[1]); Gaisberg, E. von (1922); Goebel, K. (1922[1]); Bugnon, P. (1924).

One of the first botanists to accept the phyllode theory was Martius[1], who, in 1835, analysed the terminology which it involved. He suggested the title "Stielblatt" or "Steleophyllum" for the linear type of Monocotyledonous leaf, since he considered that the word "phyllode" was better confined to those cases in which the leaf has undergone a less irrevocable reduction—that is to say, to those leaves in which the comparison of related forms, or the occasional occurrence of a rudimentary lamina, at once demonstrates that we are dealing with a metamorphosed petiole. This distinction seems to have a real value, and it is regrettable that there is no equivalent expression in our language so concise and significant as "Stielblatt." I am afraid that, if any English term—such as "petiolar leaf"—be coined, its cumbrousness and ambiguity will prevent its general adoption as a substitute for the word "phyllode" in describing Monocotyledonous leaves.

Among Monocotyledons there are certain leaves which, both in appearance and structure, so closely suggest leaf-stalks that their interpretation as petiolar phyllodes becomes an easy and natural matter. Such an organ as the leaf of *Triglochin maritima*, with its basal sheath and awl-like "radial" limb, comes into this category (fig. xxxi, A, p. 55 and fig. lxxxiii, 10, p. 110). The cylindrical leaves of certain species of *Allium* (fig. lxxvi, 23 C and 24 B, p. 101) are also petiolar, as far as appearance and anatomy are concerned[2]. And, moreover, we can find a structural parallel for the limb of *Allium fistulosum*, L. (fig. lxxvi, 23 D) in the petiolar phyllode of a certain *Acacia* (*A. leptospermoides*, Benth., fig. lxxvi, 27), which is unusual in the genus in being flattened in the horizontal plane. In this connexion, the leaves of the Iridaceae are of special interest, for they include a whole series of leaf types for which parallels can be traced among *Acacia* phyllodes. The cylindrical limbs of *Iris xiphioides*, Ehr. (fig. lxxvii, 31 D, p. 102) and of *Sisyrinchium junceum*, E. Mey. (fig. lxxviii, 27, p. 103) may be compared with the cylindrical phyllode of *Acacia teretifolia*, Benth. (fig. lxxviii, 28). And even such an unusual structure as the tetragonal limb of *Gladiolus ornatus*, Klatt (fig. lxxix, 45 B) finds its peculiarities repeated in the phyllode of another *Acacia* (*A. incurva*, Benth., fig. lxxix, 46). Every transition can, indeed, be traced between cylindrical leaves and those in which the limb is flattened in a vertical plane, giving a sword-like form. Such leaves, which are the commonest type among the Irids, are described as "isobilateral and equitant." Like the cylindrical leaves, they recall—even in the minor details of structure—the phyllodes of many Acacias. This parallelism is traced in figs. lxxix, 41—44, which represent the leaves of certain species of *Gladiolus*, and the phyllodes of those Acacias with which they may be compared. The Acacias in question (*A. uncinella*, Benth. and *A. neurophylla*, W. V. Fitz.) share with these Irids the peculiarity that the median bundle tends to insignificance,

---

[1] Martius, C. F. P. von (1835), p. 22.   [2] This view is confirmed by Menz, G. (1922).

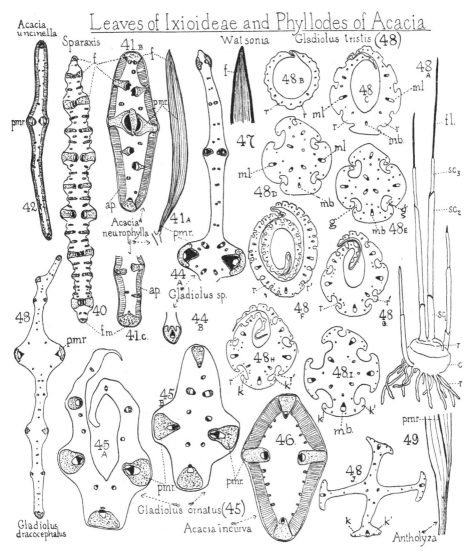

Leaves of Ixioideae and Phyllodes of Acacia (48)

Fig. lxxix. (Throughout, *p.m.r.*, pseudo-midrib; *f.*, fibres; *f.m.*, fibrous margin.) Fig. 40, *Sparaxis pulcherrima*, Hook. (× 14). Fig. 41, *Acacia neurophylla*, W. V. Fitz.; fig. 41 A, phyllode (× ½); fig. 41 B, transverse section phyllode not far from base, two upper bundles not yet united (× 14); *a.p.*, assimilating parenchyma; fig. 41 C, margin of a similar phyllode (whether dorsal or ventral uncertain) to show great development of fibres in proportion to vascular tissue. Fig. 42, *Acacia uncinella*, Benth., phyllode (× 14). Fig. 43, *Gladiolus dracocephalus*, Hook. (× 14). Fig. 44, *Gladiolus* sp.; fig. 44 A, adaxial side of limb to show fibrosis of pseudo-midrib, *p.m.r.*, and of ventral margin (× 14); fig. 44 B, dorsal margin of sheath-region to show fibrosis (× 47). Fig 45, *Gladiolus ornatus*, Klatt (× 47); fig. 45 A, sheath; fig. 45 B, limb. Fig. 46, *Acacia incurva*, Benth., transverse section phyllode (× 23) for comparison with fig. 45. Fig. 47, apex of leaf of *Watsonia marginata*, Ker-Gawl.; fibrous rim, *f.*, which is bright yellow in herbarium material, indicated in black (× ½). Fig. 48, *Gladiolus tristis*, L.; fig. 48 A, young plant (× ½); sc_1, sc_2, sc_3, successive scale leaves; *f.l.*, foliage leaf; *c.*, corm; *r.*, roots. (Brown scale leaves have been removed from corm.) Figs. 48 B—E, series of transverse sections from base upwards through tallest leaf in fig. 48 A (× 23); *r.*, subsidiary ridges; *g.*, grooves; *m.l.*, main lateral bundles; figs. 48 F—J, similar series through another leaf (× 14); *k.* and *k'.*, keels; *r.* and *r'.*, subsidiary ridges. Fig. 49, *Antholyza nervosa*, Thunb., junction of top of sheath and base of limb (× ½). [Arber, A. (1921[1]).]

while the two principal lateral veins are prominent and form a pseudo-midrib (*p.m.r.*, figs. lxxix, 41—43). The comparison between Irid leaves and *Acacia* phyllodes can be followed farther in fig. lxxx, which illustrates the structure of a number of examples in which no definite pseudo-midrib is developed.

The interest of the isobilateral equitant leaf, and of its resemblance to the petiolar phyllode of *Acacia*, is enhanced when we realise that the type in question is not confined to the Iridaceae, but that it recurs again and again throughout the Monocotyledons, though it is apparently never found among normal Dicotyledonous leaves. Isobilateral equitant leaves are known from the Helobieae (Juncaginaceae, fig. lxxxiv, 4, p. 111)[1], Liliiflorae (fig. xxxviii, p. 62, and fig. lxxx, Liliaceae, Amaryllidaceae, Iridaceae), Farinosae[2] (Restionaceae, fig. lxxxi, 3, p. 108, Xyridaceae, fig. lxvi, p. 88, Philydraceae, fig. lviii, 22, p. 79), Spathiflorae (Araceae, fig. xxxviii, 18, p. 62) and Microspermae (Orchidaceae, figs. clx, 4 and 6, p. 229). These truly ensiform leaves must be distinguished from the pseudo-ensiform leaves of *Phormium* and *Dianella*, which seem to have approximated to the isobilateral type by secondary fusion (fig. lxxxii, p. 109).

It is possible that—even if the petiolar theory be accepted for "radial" and ensiform leaves, and for those leaves whose form and anatomy suggest derivation from the radial type by flattening in the horizontal plane (fig. lxxxiii, p. 110)—a difficulty may still be felt in applying this interpretation to the thin ribbon-like leaves with a single series of bundles, which are so frequent among Monocotyledons. This difficulty vanishes, however, when we look into the matter more closely. For in the Helobieae, where ribbon-leaves often occur, the essential identity of the solid petiole-like leaves with leaves of the ribbon type is readily demonstrable[3]. For instance, within the genus *Cymodocea* (Potamogetonaceae) we find not only typical ribbon-leaves in *C. nodosa*, Aschers. (fig. lxxxiv, 12, p. 111) and typical petiolar leaves of "radial" structure in *C. isoetifolia*, Aschers. (fig. 11), but also an intermediate type in *C. manatorum*, Aschers. (fig. 10). The ribbon-leaf of *Sagittaria sagittifolia*, L. (Alismaceae, fig. lxxxiv, 3) is, again, clearly equivalent to the awl-like leaf of the Sagittarias of the *teres* group (fig. lxxxiv, 1). And, among the Juncaginaceae, the genus *Triglochin* furnishes examples of typical "radial" leaves, such as those of *T. maritima*, L. (fig. lxxxiii, 10, p. 110), and also of well-developed ribbon-limbs, in the case of *T. procera*, R.Br. These examples leave little doubt that, if the radial leaves of *Triglochin*, *Allium*, etc., may be explained as petiolar phyllodes, the same interpretation must of necessity be extended to corresponding leaves in which the limb has a ribbon form. And, further, it must be remembered that, just as Dicotyledonous petioles may be characterised either by a more or less complete ring, or else by an open arc of bundles, so petiolar phyllodes are

[1] Arber, A. (1921[2]).   [2] *Ibid.* (1922[3]).   [3] *Ibid.* (1921[2]) and (1924[2]).

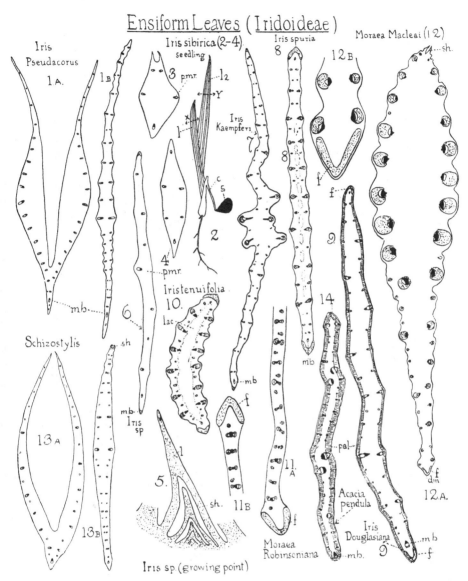

Ensiform Leaves (Iridoideae)

Iris Pseudacorus 1 A. 1 B — Iris sibirica (2-4) seedling 3 pmr 12 — Iris Kaempferi — 2 — 4 pmr — Iris spuria 8 8 — Moraea Macleai (12) 12 B sh. — Iris tenuifolia 10. lac. — 14 mb — Schizostylis sh — 6 — mb — mb Iris sp — 13 A — 13 B — 5. sh. 11 B — Iris sp. (growing point) — 11. A — Moraea Robinsoniana mb. — Acacia pendula — Iris Douglasiana mb 9 f — 12 A. dm

Fig. lxxx. (All figures, unless otherwise stated, are drawings of transverse sections of limbs of leaves, with the adaxial side towards the top of the page; xylem, solid black; phloem, white; fibres, dotted; *m.b.*, median bundle.) Fig. 1, *Iris Pseudacorus*, L. (×7), only principal bundles indicated; fig. 1 A, sheath; fig. 1 B, limb. Figs. 2—4, *Iris sibirica*, L.; fig. 2, seedling (nat. size); *s.*, seed; *c.*, cotyledon; l_1 and l_2, first and second plumular leaves. Fig. 3, transverse section of l_1 at X (×23). Fig. 4, transverse section of second leaf of another seedling at about level Y in l_2 (×23). Fig. 5, *Iris* sp., longitudinal section of stem apex with young leaves; *l.*, solid limb; *sh.*, sheath (×14). Fig. 6, *Iris* sp., China (×14). Fig. 7, *Iris Kaempferi*, Sieb. (×7). Fig. 8, *Iris spuria*, L. (×14). Fig. 9, *Iris Douglasiana*, Herb. (×14), the palisade parenchyma, *pal.*, more developed on left-hand face. Fig. 10, *Iris tenuifolia*, Pall. (×14); *lac.*, lacuna crossed by trabeculae. Fig. 11, *Moraea Robinsoniana*, C. Moore et F. Muell. (×14); to economize space the section is drawn in two parts, A and B; *f.*, fibres. Fig. 12, *Moraea Macleai*, Hort.; fig. 12 A, limb at extreme top of junction with sheath, *sh.*; *d.m.*, dorsal margin; *f.*, fibres (×14); fig. 12 B, dorsal margin of fig. 12 A (×47). Fig. 13, *Schizostylis coccinea*, Backh. et Harv. (×7); fig. 13 A, sheath; fig. 13 B, limb. Fig. 14, *Acacia pendula*, A. Cunn. (×14). [Arber, A. (1921[1]).]

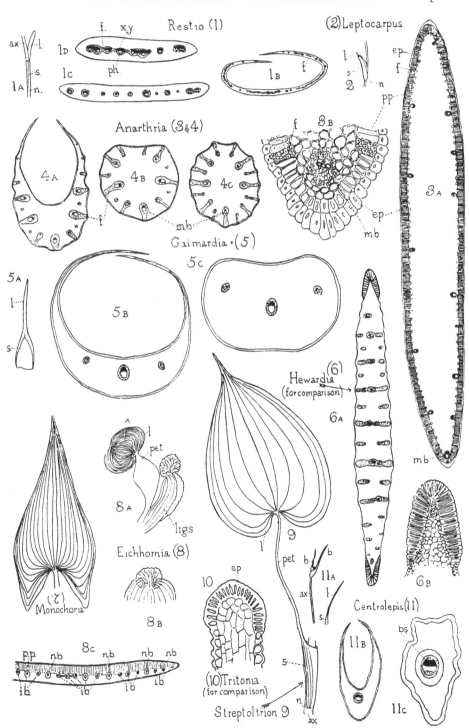

Restio (1)

(2) Leptocarpus

Anarthria (3 & 4)

Gaimardia · (5)

Hewardia (6)
(for comparison)

Eichhornia (8)

Monochoria

Tritonia
(for comparison)

Streptolirion 9

Centrolepis (11)

Fig. lxxxi. Leaves of Farinosae. Fig. 1 A—D, *Restio tremulus*, R.Br.: A, axis bearing leaf ($\times \frac{1}{2}$); B, trans. sect. of sheath ($\times 14$); C and D, transverse section of limb of another leaf ($\times 23$); D, nearer apex than C. Fig. 2, *Leptocarpus peronatus*, Mast., axis bearing leaf with sheath, *s.*, and reduced limb, *l.* ($\times \frac{1}{2}$). Fig. 3A, B, *Anarthria scabra*, R.Br.; A, transverse section of leaf-limb ($\times 14$); B, dorsal (abaxial) margin of A ($\times 77$). Fig. 4 A—C, *Anarthria gracilis*, R.Br., transverse section of one leaf: A, sheath (section slightly reconstructed at margins of sheath); B, basal part of limb; C, limb ($\times 23$). Fig. 5 A—C, *Gaimardia australis*, Gaudich.; A, leaf ($\times 3\frac{1}{2}$, *circa*); B and C, transverse sections of one leaf ($\times 77$); B, sheath; C, limb. Fig. 6 A, B, *Hewardia tasmanica*, Hook. (for comparison); A, transverse section of leaf-limb ($\times 23$); B, upper margin of A ($\times 77$). Fig. 7, *Monochoria hastaefolia*, Presl, limb of leaf to show venation ($\times \frac{1}{2}$). Fig. 8 A—C, *Eichhornia speciosa*, Kth.; A, small leaf to show dilated petiole and ligular sheath (*lig.s.*) ($\times \frac{1}{2}$); B, top of sheath from A, viewed from adaxial side to show three ligular lobes ($\times \frac{1}{2}$); C, transverse section through margin of leaf limb in A, in direction of arrow ($\times 14$); *n.b.*, normally orientated bundles; *i.b.*, inverted bundles. Fig. 9, *Streptolirion volubile*, Edgw., axis bearing leaf ($\times \frac{1}{2}$). Fig. 10, *Tritonia*, garden hybrid (for comparison), margin of ensiform leaf ($\times 193$). Fig. 11 A—B, *Centrolepis aristata*, R. and S.: A, axis of small plant bearing inflorescence inclosed in bracts (*b,b*), foliage-leaf (*l*) to right ($\times \frac{1}{2}$); B, transverse section of sheath of foliage-leaf ($\times 23$); C, transverse section of limb ($\times 77$); *b.s.*, bundle sheath, consisting of one inner thick-walled layer, and one outer layer of larger cells with thinner walls. [Arber, A. (1922^5).]

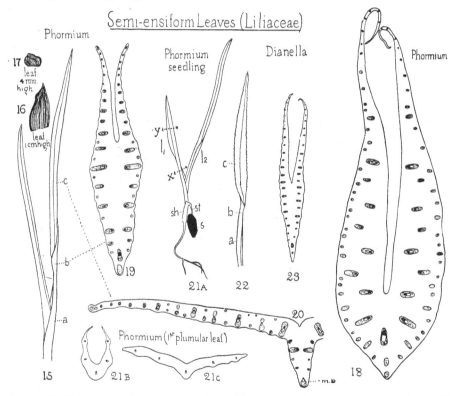

Fig. lxxxii. Figs. 15—21, *Phormium tenax*, Forst. Fig. 15, two of the younger leaves of a shoot ($\times \frac{1}{4}$). The left-hand leaf is open to the base, with no concrescent region; the right-hand leaf has an open sheath at *a*, a concrescent region at *b*, and a flat limb at *c*. Fig. 16, youngest leaf of shoot in fig. 15, open to base (slightly enlarged). Fig. 17, leaf 4 mm. high from another shoot, open to base (slightly reduced). Fig. 18, transverse section at *a* across sheathing region of right-hand leaf in fig. 15 ($\times 9$, *circa*). Fig. 19, transverse section at *b* across concrescent region of right-hand leaf in fig. 15 ($\times 9$, *circa*). Fig. 20, transverse section at *c* (incomplete) across limb region of right-hand leaf in fig. 15 ($\times 9$, *circa*). Figs. 21 A—C, seedling; fig. 21 A, seedling ($\times \frac{2}{3}$); l_1, 1st foliage-leaf; l_2, 2nd leaf; *sh.*, sheath, and *st.*, stalk of cotyledon; *s.*, seed; fig. 21 B, transverse section of l_1 in fig. 21 A, at level X ($\times 15$, *circa*); fig. 21 C, transverse section of l_1 in fig. 21 A at level Y ($\times 15$, *circa*). Figs. 22 and 23, *Dianella nemorosa*, Lam. Fig 22, leaf with sheathing region, *a*, concrescent region, *b*, and open limb, *c* (reduced). Fig. 23, transverse section of leaf in fig. 22 at level *b*, where fusion is most complete ($\times 9$, *circa*). [Arber, A. (1921^1).]

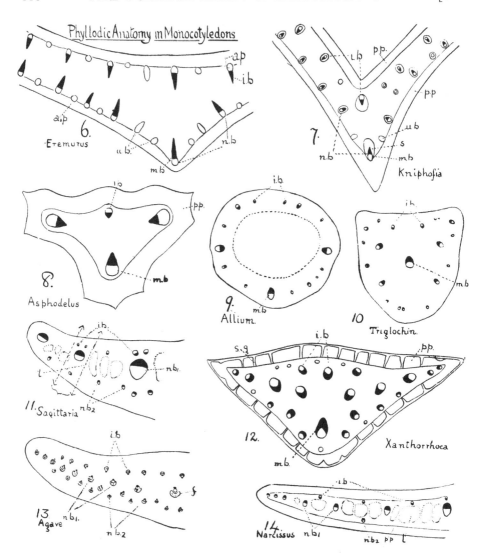

Fig. lxxxiii. (Lettering throughout as follows: *m.b.*, median bundle; *n.b.*, normal bundle; *n.b.*1, normal bundle of main series; *n.b.*2, normal bundle of second series; *i.b.*, inverted bundle; *u.b.*, undifferentiated bundle; *f.*, fibres; *l.*, lacuna; *p.p.*, palisade parenchyma; *a.p.*, assimilating parenchyma. Xylem throughout represented solid black, phloem white, and fibres dotted.) Fig. 6, *Eremurus himalaicus*, Baker, transverse section of midrib and adjacent region of very young leaf (× 14). Fig. 7, *Kniphofia caulescens*, Baker, transverse section of midrib region of leaf (× 14). Fig. 8, *Asphodelus luteus*, L., transverse section of leaf (× 47). Fig. 9, *Allium cepa*, L., transverse section of upper part of leaf (× 14). Fig. 10, *Triglochin maritima*, L., transverse section of leaf (× 23). Fig. 11, *Sagittaria montevidensis*, Cham. et Schlecht, transverse section of part of leaf including margin (× 23). The part enclosed between the arrows is shown in detail in fig. lvii, 32, p. 78. Fig. 12, *Xanthorrhoea* sp., transverse section of leaf (× 14), hypoderm and girders (*s.g.*) sclerised. Fig. 13, *Agave densiflora*, Hook., transverse section of part of leaf, including margin (× 14). Fig. 14, *Narcissus pseudo-narcissus*, L., transverse section of part of leaf, including margin (× 23). [Arber, A. (1918[2]).]

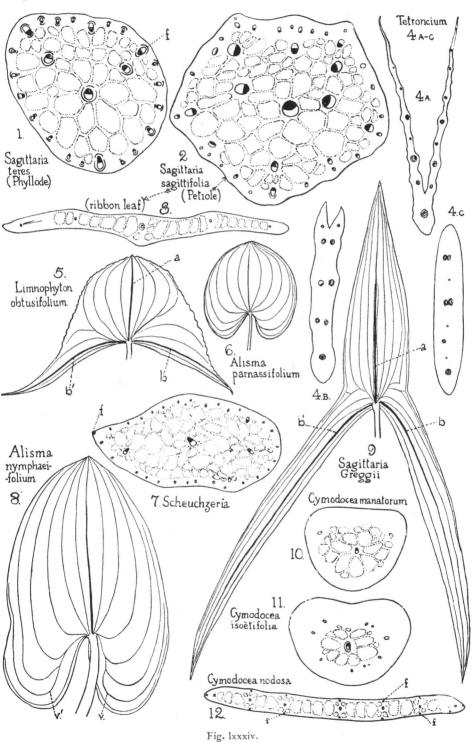

1. Sagittaria teres (Phyllode)

2. Sagittaria sagittifolia (Petiole)

3. (ribbon leaf)

4 A-C Tetroncium

4.A.

4.c

4.B.

5. Limnophyton obtusifolium

6. Alisma parnassifolium

7. Scheuchzeria

8. Alisma nymphaei-folium

9. Sagittaria Greggii

10. Cymodocea manatorum

11. Cymodocea isoëtifolia

12. Cymodocea nodosa

Fig. lxxxiv.

Fig. lxxxiv. Leaves of Helobieae. In this figure xylem shown in black, phloem in white, fibres (*f*.) dotted, and the outlines of lacunae in dotted lines. Fig. 1, *Sagittaria* of *S. teres* group: transverse section of limb of leaf; *f.*, interrupted fibrous sheath of bundle (slight asymmetry of section probably due to incomplete recovery of herbarium material used; Georgia Plants. Roland Harper 1473. Ex Herb. Brit. Mus.) (× 23). Fig. 2, *Sagittaria sagittifolia*, L.: transverse section of petiole close to blade, fibrous bundle-sheath (less highly developed than in species shown in fig. 1) not indicated (× 14). Fig. 3, *Sagittaria sagittifolia*, L.: transverse section of small ribbon leaf (× 23). Fig. 4, *Tetroncium magellanicum*, Willd.: transverse sections of leaf; A, sheath; B, transition to limb; C, limb (× 23). Fig. 5, *Limnophyton obtusifolium*, Miq.: blade of leaf; *a*, midrib; *b*, *b'*, cusp veins (serration of margin probably an effect of drying) (× ⅓). Fig. 6, *Alisma parnassifolium*, Bassi, var. *majus*, blade of leaf (× ½). Fig. 7, *Scheuchzeria palustris*, L.: transverse section of limb of leaf; *f*, fibrous strand occupying one margin (on account of small scale, fibrous sheaths of bundles not separately indicated) (× 23). Fig. 8, *Alisma nymphaeifolium*, Griseb.: blade of leaf; *v*, *v'*, principal veins of auricles (× ½). Fig. 9, *Sagittaria Greggii*, Smith.: blade of leaf; *a*, midrib; *b*, *b'*, cusp veins (× ½). Fig. 10, *Cymodocea manatorum*, Aschers.: transverse section of limb of leaf (× 23). Fig. 11, *Cymodocea isoetifolia*, Aschers.: transverse section of limb of leaf; xylem indistinguishable in smaller bundles surrounding median bundle (× 23). Fig. 12, *Cymodocea nodosa*, Aschers.: transverse section of limb of leaf (× 23). [Arber, A. (1921²).]

not necessarily of the "radial" type to which we are accustomed in certain Acacias. *Oxalis bupleurifolia*, A. St. Hil. (figs. lxxxv, 3 A and 3 B) and, possibly *Lathyrus Nissolia*, L. (fig. lxxv, 5, p. 99) are examples of plants bearing foliar organs equivalent to petioles, but expanded in the horizontal plane, and traversed by a single series of bundles. In this connexion it may be noted that there is no hard-and-fast line between the anatomical characters of petioles and leaf-sheaths, for while petioles may resemble leaf-sheaths in having a single series of bundles with their xylems directed upwards, leaf-sheaths occasionally have some bundles towards their upper surface, orientated in an inverse sense as compared with the main series— thus repeating a character which is extremely common in petioles, but not typical of leaf-bases (e.g. *Thalictrum*[1], fig. lxxxvi, 12, p. 114).

Up to this point we have considered cases in which there is both a sheathing leaf-base and a distinct limb, but in which the limb is the more conspicuous region. Expressing the matter in terms of the phyllode theory, we should say that these leaves consist of leaf-base and petiole alone, the petiole forming the principal part of the leaf. But the proportion which these two elements bear to one another is subject to wide variation. In *Iris xiphioides*, for instance, the cylindrical petiolar limb may be of some length, or it may, on the other hand, be reduced to a mere trace, so that the leaf consists almost exclusively of sheath, the vestigial petiole forming merely a short cylindrical apex (cp. figs. lxxvii, 31 A and B, p. 102). These latter proportions have become stereotyped in certain members of the Liliaceae (e.g. *Hyacinthus*, figs. lxxxvii, 2 A and B, p. 115) in which the chief part of the leaf is of leaf-sheath nature, while the petiole only contributes a solid tip, containing a closed ring of bundles. We find analogies for such leaves—both as regards form and anatomy—in the bud-scales of the Dicotyledonous *Fatsia japonica*, Decne., some of which have cylindrical apices, undoubtedly corresponding to the stalks of the normal foliage-leaves (figs. lxxxvii, I A, B, C). Among the

[1] Worsdell, W. C. (1908).

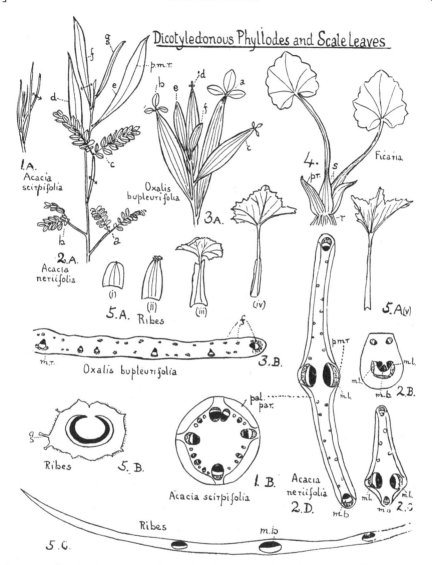

Fig. lxxxv. (In all diagrams of sections, xylem is represented black, phloem white, and fibres, f., dotted.) Fig. 1 A and B, *Acacia scirpifolia*, Meissn. Fig. 1 A, apical region of shoot with cylindrical phyllodes (reduced); fig. 1 B, transverse section of phyllode ($\times$ 30); *pal. par.*, palisade parenchyma. Figs. 2 A—D, *Acacia neriifolia*, A. Cunn. Fig. 2 A, upper part of seedling (reduced); a and b, normal leaves; c, leaf with petiole slightly expanded; $d—g$, phyllodes; *p.m.r.*, pseudomidrib; fig. 2 B, transverse section of petiole of a leaf lower on the axis than leaf a; fig. 2 C, transverse section of petiole of leaf c; fig. 2 D, transverse section of phyllode f; *p.m.r.*, pseudomidrib, derived from the two main laterals (*m.l.*). Figs 2 B—2 D ($\times$ 30). Fig. 3 A and B, *Oxalis bupleurifolia*, A. St. Hil. Fig. 3 A, apical region of shoot (reduced). The successive leaves, $a—d$, show progressive reduction of lamina, which in e and f is entirely lost. Fig. 3 B, transverse section of part of phyllode including midrib, *m.r.* ($\times$ 30). Fig. 4, *Ficaria verna*, Huds., part of basal region of plant to show two reduced leaves (*pr.*) corresponding to sheathing bases (*s*) of normal leaves. Figs. 5 A—C, *Ribes nigrum*, L. Fig. 5 A (i)—(v), successive leaves of bud, showing transitions from bud-scale to normal leaf (reduced); fig. 5 B, transverse section of petiole; g, glandular emergence ($\times$ 17); fig. 5 C, transverse section of bud-scale (incomplete); *m.b.*, midrib ($\times$ 17). [Arber, A. (1918[2]).]

Monocotyledons, leaf-base phyllodes with petiolar apices are by no means
confined to the Liliaceae—*Doryanthes* (Amaryllidaceae, fig. lxxxviii, 1, p. 116),

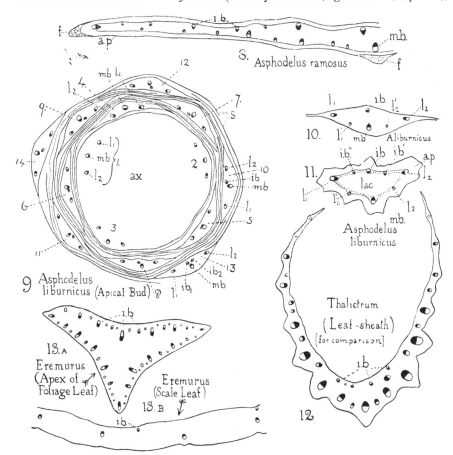

Fig. lxxxvi. (Xylem, black; phloem, white; fibres, dotted.) Fig. 8, *Asphodelus ramosus*, L.; transverse section of half a leaf, including median bundle (*m.b.*); *f.*, fibres; *a.p.*, assimilating parenchyma;
i.b., inverted bundle. (This section was from herbarium material, which possibly had not recovered
its normal thickness) (× 14). Figs. 9—11, *Asphodelus liburnicus*, Scop. Fig. 9, transverse section
near apex of axis, *ax.*, showing a number of young leaves (1—14) with divergence $\frac{5}{13}$ (× 14). In
each leaf, *m.b.*, median bundle; l_1 and l_2, lateral bundles; *i.b.*, inverted bundle derived from
median bundle. In leaf 13, two bundles, *i.b_1* and *i.b_2*, are derived from the median bundle. *s.*,
sheathing wings of leaf-bases. Fig. 10, transverse section through another leaf cut at a higher level,
showing l_1' and l_2', which have been given off from l_1 and l_2 (× 14). Fig. 11, transverse section,
higher still in the limb of another leaf, showing *i.b.'* and *i.b.''*, which have been given off from *i.b.*;
lac., lacuna; *a.p.*, assimilating parenchyma. Fig. 12, *Thalictrum flavum*, L.; transverse section
of leaf-sheath to show inverted bundles, *i.b.* (× 14). Fig. 13, *Eremurus himalaicus*, Baker. Fig.
13 A, transverse section near apex of foliage leaf (× 14); fig. 13 B, part of transverse section of
scale-leaf (× 14); *i.b.*, inverted bundle. [Arber, A. (1920³).]

Distichia (Juncaceae, fig. 6) and *Elegia* (Restionaceae, fig. 5) afford other
examples. In this connexion it is significant that Hallier[1], in discussing

[1] Hallier, H. (1912).

the parts of the Dicotyledonous flower, has interpreted the "cornet" or "aiguilla," terminating certain sepals, as a vestigial petiole, while he considers the main part of the sepal as a leaf-sheath. On this interpretation, the examples that he gives, such as the sepal of *Passiflora* (fig. lxxxviii, 10, p. 116), correspond precisely to the Monocotyledonous foliage leaves which we have been considering (cp. fig. lxxxviii, 10 A, with 1, 2 A, 5 A, 6 A).

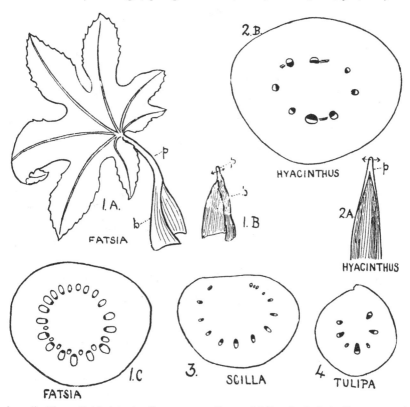

Fig. lxxxvii. Fig. 1, *Fatsia japonica*, Decne.: A, small normal foliage leaf; *b.*, leaf-base; *p.*, petiole; B, bud-scale; *b.*, leaf-base; *p.*, rudiment of petiole; c, transverse section of apex of bud-scale at position marked with arrow in B; A and B, (half natural size); c (× 23); fig. 2, *Hyacinthus* (garden var.): A, apex of leaf (half natural size); B, transverse section through apex of leaf shown in A, at level of arrow (× 23); fig. 3, *Scilla* (garden var.): transverse section through apex of leaf which was flat and dorsiventral except at tip (× 14); fig. 4, *Tulipa sylvestris*, L., transverse section through apex of leaf which was flat and dorsiventral except at tip; form on upper side shows first indication of opening out into main flat part of leaf (× 23). [Arber, A. (1920[1]).]

In the genus *Asparagus*, we find curious scale-leaves, which are continued basally into a tail or spine (fig. lxxxix, 1, p. 117). Though these leaves, on account of the downward direction of the spine, look, at first glance, very different from the leaves shown in fig. lxxxviii, yet, on close examination, they are found to conform to the same type. The spine is not a mere emergence, but is the only part of the leaf which receives its vascular supply

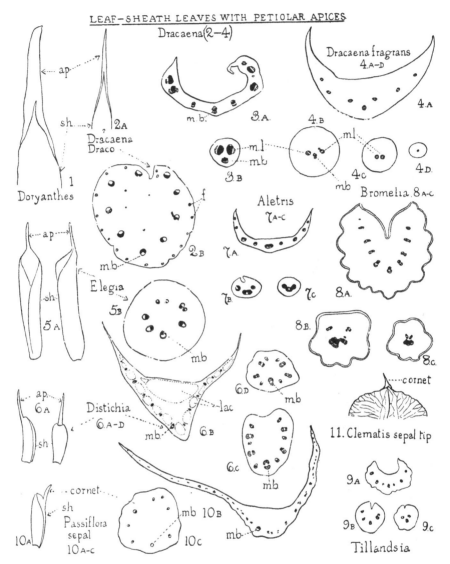

LEAF-SHEATH LEAVES WITH PETIOLAR APICES
Dracaena (2–4)

Fig. lxxxviii. (Xylem, black; phloem, white; fibres, dotted; *m.b.*, median bundle; *m.l.*, main lateral bundle; *lac.*, lacuna; *sh.*, leaf-sheath region; *ap.*, petiolar apex.) Fig. 1, tip of leaf of *Doryanthes Guilfoylei*, W. M. Bailey (nat. size). Fig. 2, *Dracaena Draco*, L.; 2 A, tip of leaf (nat. size); 2 B, transverse section at base of apex (× 14); *f.*, strands of fibres. Fig. 3, *Dracaena* "Duchess of York"; 3 A, transverse section of leaf near tip; 3 B, transverse section of solid apex to show importance of main laterals (× 14). Fig. 4, *Dracaena fragrans*, Ker-Gawl., series of transverse sections through tip of leaf (× 14); fig. 4 D is from a second leaf. Fig. 5, *Elegia deusta*, Kunth; 5 A, two views of leaf (nat. size); 5 B, transverse section of apex (× 23). Fig. 6, *Distichia clandestina*, Buchen.; 6 A, two views of leaf (nat. size); 6 B—D, series of transverse sections through upper part of leaf (× 23); in the herbarium material used, xylem and phloem were not distinguishable. Fig 7, *Aletris farinosa*, L., series of transverse sections through tip of leaf (× 14). Fig. 8, *Bromelia* "*macrodonta*" (=? *Ananas macrodontes*, E. Morren), series of transverse sections through tip of leaf (× 14), margin fibrous. Fig. 9, *Tillandsia Lescaillei*, Wright, series of transverse sections through leaf-tip (× 14). Fig. 10, *Passiflora incarnata*, L.; 10 A, sepal (nat. size); 10 B, transverse section of sheath region (× 14); the asymmetry is natural; the bundles are nearly all cut obliquely; 10 C, transverse section of "cornet" (× 23), *m.b.*, ? median bundle. Fig. 11, *Clematis* sp., apex of sepal from ventral side to show "cornet" (nat. size). [Arber, A. (1922[1]).]

direct from the axis (figs. lxxxix, 2 and 3); the upper scale-like region of the leaf is entered exclusively by branches (*l.b.*), arising from the spine-bundle (*s.b.*). These facts seem to me to support the view, first suggested by Buscalioni[1], that the scale is a ligular sheath, while the spine is the petiolar limb.

We have seen that in Monocotyledonous leaves, the petiole may be reduced to a mere inconspicuous apex, but reduction does not always stop. at this point, and sometimes no region survives except the leaf-base. For this degree of reduction, also, we can find analogies among Dicotyledons. Fig. lxxv, 2 A—C, p. 99, for example, shows the leaves of *Petasites vulgaris*, Desf., whose laminae, in normal cases, may be three feet across, with a petiole and a relatively inconspicuous sheath. But, in association with the

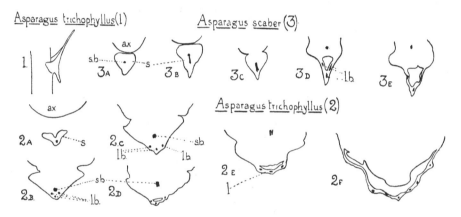

Fig. lxxxix. Spinous scale-leaves of *Asparagus*. Figs. 1 and 2, *A. trichophyllus*, Bunge; fig. 1, small segment of axis bearing scale-leaf, to show tail or spine; structures subtended by the leaf removed (enlarged); figs. 2 A—F, transverse sections through base of leaf with spine, *s.*, showing its vascular relation to the axis, *ax.*, and the upper part of the leaf, *l.*; *s.b.*, spine bundle; *l.b.*, bundles for upper part of leaf (× 14). Figs. 3 A—E, *A. scaber*, Brign., series of sections through leaf-base, corresponding to those in fig. 2 and lettered similarly (× 14). [A.A.]

inflorescences, we find leaves in which the lamina and petiole are gradually reduced in proportion to the sheath, until finally only the sheath region remains (fig. lxxv, 2 C). The foliage-leaves of such Monocotyledons as the Day-lily, *Hemerocallis*[2], probably fall into the same leaf-base category as the reduced inflorescence-leaves of *Petasites vulgaris*.

We thus take a leaf-base terminating in a simple petiole as the fundamental Monocotyledonous type, and regard leaves such as those of *Hemerocallis* as representing a further degree of reduction, in which the petiole is altogether lost. But it is possible for the course of evolution to proceed in quite another direction, in which the trend is towards the elaboration of the petiole, rather than towards its reduction. The various leaf forms met with in

---

[1] Buscalioni, L. (1920—21).   [2] Arber, A. (1920[1]).

the Palms[1] and Irids[2] are particularly useful in helping us to understand the process. Figs. xc and xci represent sections from series taken from below

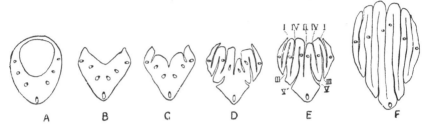

Fig. xc. *Trachycarpus Fortunei*, H. Wendl. Series of transverse sections from below upwards through the second plumular leaf of seedling (× 23). (Sections cut by Ethel Sargant.) Fig. A, sheath; fig. B, petiole; figs. C—E, process of invagination; fig. F, final form. In fig. E, the grooves are numbered in the order of their appearance. In this figure only principal bundles included. [Arber, A. (1922[2]).]

upwards through young plumular leaves of two Palms, belonging to the genera *Trachycarpus* and *Cocos*. It will be seen that the sheathing base is

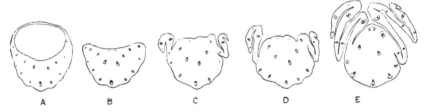

Fig. xci. *Cocos Romanzoffiana*, Cham. Series of transverse sections from below upwards through second foliage leaf of seedling (× 14). Fig. A, sheath; fig. B, petiole; figs. C—E, stages in invagination. [Arber, A. (1922[2]).]

in each case succeeded by a short petiolar region, B. It is the following sections (C, D, etc.) which show the mechanism by which the petiole is elaborated into a "pseudo-lamina." A series of invaginations (numbered in order of their occurrence, I—V, in fig. xc, E) penetrate into the tissues between the vascular bundles, and result in a form of leaf rudiment which has often been inaccurately described as "folded." This fan-like body ultimately expands, and

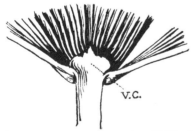

Fig. xcii. *Trachycarpus excelsus*, H. Wendl. Ventral view of junction of leaf-limb and petiole to show "ligule" or ventral crest, *v.c.* (× ½). [Arber, A. (1922[2]).]

by a process of secondary splitting, develops into the distinctive leaf of the Palm. But the "fan" or "feather" thus produced—which sometimes attains gigantic dimensions—is, on the present interpretation, not a true blade equivalent to that of the Dicotyledonous

[1] Arber, A. (1922[2]).   [2] *Ibid.* (1921[1]).

leaf, but, as has just been shown, a mere elaboration of the distal region of the petiole. The so-called "ligule", which occurs at the base of the blade in Fan-palms (fig. xcii), is, on my view, not a ligule at all, but merely the ventral surface of the proximal part of the petiole, under which the invaginations, which affect the more distal region, have, as it were, burrowed, so as to leave an overarching penthouse of tissue (fig. xciii). The "dorsal scale," which, in some Palms, reaches as high a degree of development[1] as

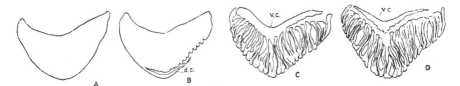

Fig. xciii. *Rhapis humilis*, Blume. Series of transverse sections through young leaf of mature plant passing through junction of petiole and leaf-limb from below upwards (× 9, *circa*); A, top of petiole; in B, invagination is just beginning and the slightly developed "dorsal scale" or dorsal crest (*d.c.*) is distinguishable; in C, the "ligule" or ventral crest (*v.c.*) is just becoming detached, while in D, it is entirely free. The asymmetry of B and C is due to a slight obliquity in the series of sections. [Arber, A. (1922²).]

the "ligule," can be interpreted on similar lines, while the solid tips of the plumular leaves of certain members of the family may also, I think, be explained as due to the dying out of the invaginations before the extreme apex is reached (fig. xciv).

Examples of invagination, as striking as those in the Palms, are to be found among the Irids. The extremely peculiar leaf of *Crocus*, for instance,

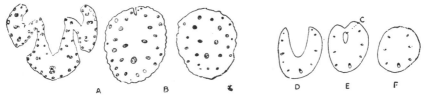

Fig. xciv. Solid tips of plumular leaves of Palms. A—C, *Phoenix dactylifera*, L. Series of transverse sections from below upwards through apex of first foliage leaf to show dying out of invaginations (× 14); bundles consist chiefly of fibres; D—E, *Pritchardia filifera*, Lind. Series of transverse sections through apex of first foliage leaf (× 14); E shows that the apex is slightly hooded; *c.*, cavity. [Arber, A. (1922²).]

arises from a solid petiole (fig. xcv, 56 C, p. 120), not by numerous grooves as in the Palms, but by one pair of invaginations, which take a spiral course (fig. xcv, 56, F—H), and thus produce two marginal wings, which, on uncoiling, are responsible for the definitive form of the leaf, with its characteristic transverse section (fig. xcv, 56 I). The curious leaf of the related genus *Romulea* (figs. xcv, 61–63), though superficially so different from that of *Crocus*, is constructed on the same fundamental plan, the only divergence

[1] Candolle, C. de (1913).

Leaves of Crocoideae

Crocus Tomasinianus (56)   Crocus vernus (57) Crocus Tomasinianus (56)

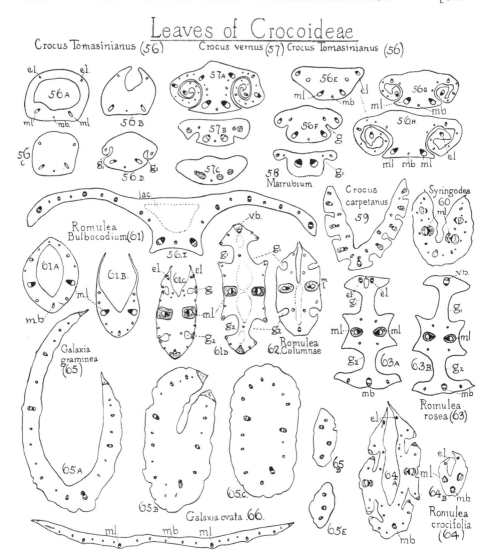

Fig. xcv. (Throughout, *m.l.*, main lateral bundle; *e.l.*, external lateral bundle ; *v.b.*, ventral bundle; g_1 and g_2, grooves; *lac.*, lacuna.) Figs. 56 A—I, *Crocus Tomasinianus*, Herb. ; figs. 56 A—D, sections from microtome series through transition from sheath to limb in first foliage-leaf of seedling (×47); figs. 56 E—I, series of transverse sections through leaf of mature plant from near base upwards, not including sheath (× 23); figs. 56 E—H, basal region; fig. 56 I, higher up. Fig. 57, *Crocus vernus*, All.; fig. 57 A, transverse section at level corresponding to fig. 56 H ; figs. 57 B and c, transverse sections near apex (× 14). Fig. 58, *Marrubium velutinum*, Sibth. et Sm. (Labiatae), transverse section of petiole (× 14). Fig. 59, *Crocus carpetanus*, Boiss. et Reut., transverse section of leaf-limb (× 23). Fig. 60, *Syringodea bicolor*, Baker, transverse section of leaf-limb (× 23). Figs. 61 A—D, *Romulea Bulbocodium*, Sebast. et Mauri, series of transverse sections through leaf (× 23). Fig. 62, *R. Columnae*, Sebast. et Mauri, transverse section just at top of sheath (× 23). Fig. 63, *R. rosea*, Eckl.; fig. 63 A, transverse section just at top of sheath; fig. 63 B, transverse section of limb (× 23). Fig. 64, *R. crocifolia*, Vis.; fig. 64 A, transverse section of leaf-limb; fig. 64 B, transverse section close to apex (× 23). Figs. 65 A—E, *Galaxia graminea*, Thunb., series of transverse sections from sheathing base to apex; possibly 65 D and E are from a different leaf (× 23). Fig. 66, *Galaxia ovata*, Thunb., transverse section of limb of leaf (× 14). [Arber, A. (1921[1]).]

being that there are four invaginations instead of two, and that the external
lateral bundles (*e.l.*), instead of being widely separated as in *Crocus*, fuse into
a single strand.

This process of invagination from a solid petiole, may give rise to leaves
perhaps more anomalous than any others in the plant world. Fig. lxxix, 48
J, p. 105, shows the cruciform transverse section of the limb of *Gladiolus tristis*
—a section which is so unlike that of a typical foliage leaf, that to explain
it without a knowledge of its history would be impossible. But in figs. lxxix,
48 B—I, which represent sections from below upwards through the junction of
sheath and limb, we can trace the origin of the final form, which is found to be
due to four deep grooves or invaginations, the outline being rendered slightly
more complex by the development of a pair of lateral keels, distally, from
each of the four arms which are left between the grooves.

Forms even more aberrant than the types just described are to be found
among those Iridaceae which possess the so-called "foliated" leaves.
Fig. xcvi, 50 N, p. 122, shows a transverse section of the "foliated" leaf-limb of
Cypella. Without the sequence of sections H—M, it would be almost impossible
to guess that this complicated form had been brought about by a series of
alternating invaginations between the principal bundles, supplemented by
wing-like outgrowths developed at the sides of these strands. A species of
Tigridia, in which the history is similar, is shown in figs. xcvi, 52 A—D,
while the relation of the different regions in the first plumular leaf of
another species—still ensheathed by the cotyledon—is shown in figs. xcvii,
21 A—D, p. 123.

In the plants which we have hitherto considered, the material, which the
invaginations have to work upon, is a more or less radial petiole. But a
petiole of the open dorsiventral type may also be modified in the same way,
as shown by the series of sections through the junction of sheath and limb of a
leaf of *Curculigo* (Amaryllidaceae) represented in figs. xcvii, 17 A—F, p. 123;
a plicate limb is here produced by alternating dorsal and ventral invaginations.

In *Veratrum* (Liliaceae, figs. xcviii, 13 A—F, p. 124), the development
follows closely similar lines. In such cases there seems to be no criterion by
which we can, with certainty, decide whether it is a dorsiventral petiole, or
the distal region of the leaf-sheath, which is subject to invagination. But this
point is probably of no great importance.

In addition to those leaves whose *ontogeny* indicates that their limbs
are elaborations of the petiole, there are other Monocotyledonous leaves
for which the petiolar interpretation is suggested, not by developmental
evidence, but by the *vascular anatomy* of the "blade." It is usual for leaf
laminae, except in the midrib and principal veins, to be supplied by a single
series of bundles, so orientated that their xylems are directed towards
the adaxial or upper surface. But I have found that, in the leaf-limbs

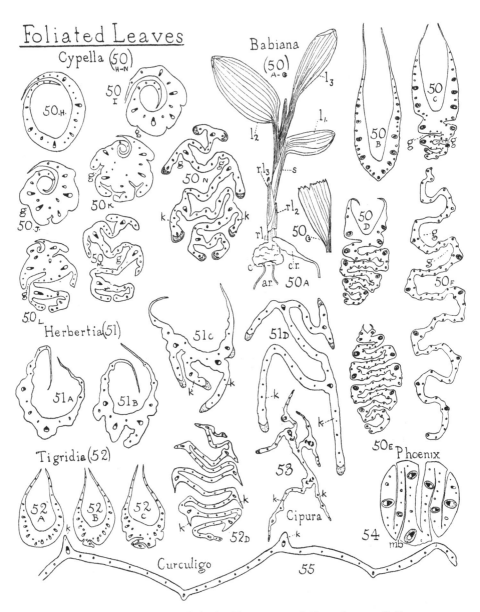

Fig. xcvi. (Throughout, *g.*, groove; *k.*, keel.) Figs. 50 A—G, *Babiana*; fig. 50 A, *Babiana* sp., young plant (× ½) ; *c.*, corm ; *a.r.*, absorptive roots ; *c.r.*, contractile root ; l_1, l_2, l_3, normal leaves ; *s.*, leaf sheath ; *r.l*$_1$, *r.l*$_2$, *r.l*$_3$, reduced leaves : figs. 50 B—F, series of transverse sections through junction of sheath and limb of l_3 in fig. 50 A (× 14) ; fig. 50 G, *Babiana cuneifolia*, Baker, limb and upper part of long leaf-sheath (× ¼) ; figs. 50 H—N, *Cypella Herberti*, Herb. (× 14) ; fig. 50 H—M, series of transverse sections through junction of sheath and limb in one leaf ; fig. 50 N, limb of another leaf. Figs. 51 A—D, *Herbertia pulchella*, Sweet, series through junction of sheath and limb (× 23). Figs. 52 A—D, *Tigridia Pavonia*, Ker-Gawl., series of transverse sections through junction of sheath and limb (× 7). Fig. 53, *Cipura paludosa*, Aubl., transverse section of limb (× 14) ; orientation uncertain, as only a fragment of the limb was available. Fig. 54, *Phoenix dactylifera*, L., transverse section near base of first foliage leaf (× 14) to show "folding." Fig. 55, *Curculigo* sp., transverse section of small part of limb near margin ; leaf can be folded up like a fan (× 14). [Arber, A. (1921[1]).]

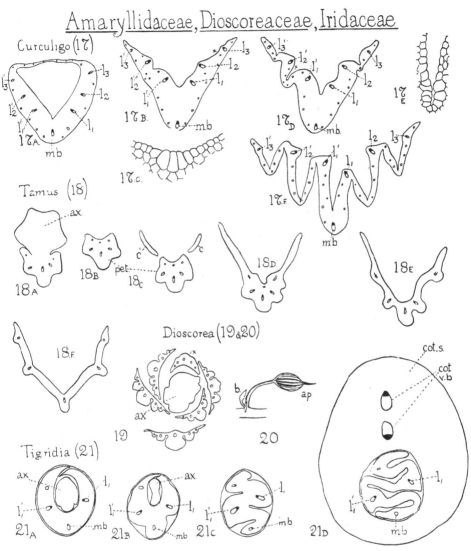

Fig. xcvii. Figs. 17 A—F, *Curculigo recurvata*, Dryand. Figs. 17 A, B, D, F, series of transverse sections through one leaf, passing from sheath, fig. 17 A, to "plicate lamina," fig. 17 F (× 14); *m.b.*, median bundle; l_1, l_2, l_3, l_1', l_2', l_3', main laterals; fig. 17 C, epidermis at region marked with cross in fig. 17 B (× 77); fig. 17 E, epidermis of base of groove marked with cross in fig. 17 D (× 77). Figs. 18 A—F, *Tamus communis*, L., series of transverse sections through one leaf (× 23) passing from attachment of leaf and axis (*ax.*, fig. 18 A) to "lamina" (fig. 18 F); *c.* and *c.'*, cordate basal lobes of "lamina" cut on either side of upper part of petiole, *pet.* Fig. 19, *Dioscorea sativa*, L., transverse section of apical bud, passing through axis, *ax.*, and a series of leaves of which the one marked with a cross is cut nearest to its level of attachment (× 14). Fig. 20, *Dioscorea* sp., young leaf with axillary bud, *b.*, to show narrow pointed apex, *ap.*, which is becoming brown and shrivelled (about nat. size). Figs. 21 A—D, *Tigridia Pringlei*, S. Wats., sections through first plumular leaf, from sheath, fig. 21 A, to plicate "lamina," fig. 21 D, from a series in Ethel Sargant's collection (× 47). The outer line represents the inner epidermis of the cotyledon sheath (*cot.s.*), which is only shown completely in fig. 21 D; *ax.*, plumular bud; *m.b.*, l_1 and l_1', main bundles of first plumular leaf; *cot.v.b.*, vascular bundle of cotyledon, which is cut twice, as it doubles on itself. [Arber, A. (1922[4]).]

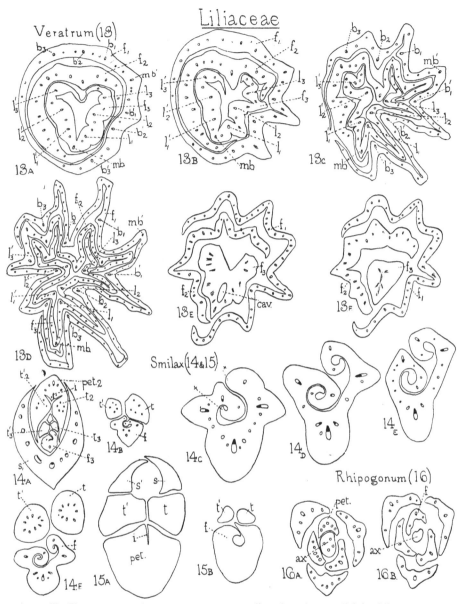

Fig. xcviii. Figs. 13 A—F, *Veratrum "album nigrum,"* sections from a slightly oblique transverse series through a shoot apex, the three innermost (f_1, f_2, and f_3) of the numerous leaves being alone represented ($\times 14$). Bundles of f_3 lettered *m.b.* (median bundle) and l_1, l_2, l_3, l_1', l_2', l_3' (lateral bundles). In Figs. 13 A, C, and D, the bundles of f_1 also are lettered *m.b.'* (median bundle), b_1, b_2, b_3, b_1', b_2', b_3' (lateral bundles). Series ranges from fig. 13 A, sheathing region of the three leaves, to fig. 13 F, which passes through the solid apex of f_3, and near the apex of f_1 and f_2; in fig. 13 E there is a cavity (*cav.*) in f_3, showing that the apex is hooded. Figs. 14 A—F, *Smilax herbacea,* L.; fig. 14 A, transverse section of apical bud ($\times 14$); the sheathing base of leaf s_1 encloses a second leaf, whose petiole, *pet.$_2$*, showing ventral invagination *i.*, is cut at the level of attachment of the tendrils, t_2 and t_2'; a third leaf is cut through the "lamina," f_3, and the two free tendrils, t_3 and t_3'; fig. 14 B, transverse section of another leaf passing through the lamina, $f.$, at more advanced stage, and tendrils, $t.$ and $t.'$ ($\times 14$); figs. 14 C—E, series of sections through developing "lamina" of another leaf ($\times 23$); the points marked with a cross in fig. 14 C will eventually become the margins of the "lamina"; fig. 14 F, section of another leaf ($\times 14$) to show developing lamina, $f.$, and tendrils $t.$ and t'. Figs. 15 A and B, *Smilax laurifolia,* L., two sections of one young leaf ($\times 47$); fig. 15 A passes through leaf at level just above sheath and shows wings of sheath, $s.$ and $s.'$, tendrils, $t.$ and $t.'$, and petiole, *pet.*, penetrated by ventral invagination, *i.*; fig. 15 B, same leaf at higher level, just below tips of tendrils; $f.$, lamina. Figs. 16 A and B, *Rhipogonum album,* R.Br., two sections from transverse series through apical bud ($\times 23$); *ax.*, axis; the petiole, *pet.*, of youngest leaf is just detached in fig. 16 A, and the same leaf is represented by the lamina, $f.$, in fig. 16 B. [Arber, A. (1922[4]).]

of the Pontederiaceae[1] and some other Monocotyledons, there are, in addition to the normally orientated bundles, a number of strands whose xylem is directed downwards. Solereder[2] had previously shown that the same is true of the Hydrocharitaceae; I include here (fig. xcix) some sketches of the leaves of a member of this family, *Trianea bogotensis*, Karst. On the phyllode theory, the peculiar orientation of the strands in the "blade," shown in figs. xcix, D and E, would be interpreted as due to its origin by the flattening and expansion of a petiole of the "radial" type, in which the bundles towards the adaxial surface are inverted as compared with those towards the lower surface. In *Trianea bogotensis* the petioles of the more strongly developed leaves are of this type (fig. xcix, C); in feebler examples

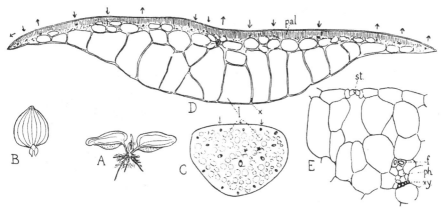

Fig. xcix. *Trianea bogotensis*, Karst. = *Limnobium stoloniferum*, Grisebach. A, small floating plant ($\times \frac{1}{2}$) to show spongy air-tissue forming cushion below leaf-blade. B, leaf from adaxial side ($\times \frac{1}{2}$) to show transparent ligulate sheath, free from the petiole in its upper region. C, transverse section of petiole of well-developed leaf to show "radial" structure ($\times 14$). D, transverse section of small leaf-limb ($\times 14$); *pal.*, palisade parenchyma; *l.*, lacunae; the arrows indicate the direction in which the xylem of the bundles points; at x a segment of a bundle is cut in a diaphragm. E, small part of upper surface of leaf in transverse section ($\times 193$) to show stomate, *st.*, and bundle with xylem, *xy.*, directed towards lower surface, while the phloem, *ph.*, and fibres, *f.*, lie above. [A.A.]

only the principal arc of bundles survives, so that the "radial" character is lost.

There seems, then, to be evidence that not only the simpler types of Monocotyledonous leaf, but also some of the more complex forms with well-defined blades, may legitimately be interpreted as mere phyllodes, elaborated by secondary differentiation from the fundamental "leaf-base + petiole" type. It remains, however, to be considered whether the phyllode interpretation can be applied generally to all Monocotyledonous leaves, or if it should be confined to the particular cases which we have discussed. There is, I think, a good deal to be said in favour of applying this interpretation generally. It has the great advantage of explaining the parallel

[1] Arber, A. (1918[2]).   [2] Solereder, H. (1913).

arrangement of the main veins, so characteristic of the limb in this class (fig. C), which would be a direct and inevitable outcome of the expansion of a petiole or rachis. Even when Monocotyledonous blades are cordate at the base, they show evident signs of derivation from a parallel-veined organ; their form can be seen to depend upon the degree of separation and outward curvature of the proximal region of the veins, which converge in their distal region and meet at the leaf-tip (e.g. *Streptolirion*, fig. lxxxi, 9, p. 108). Such blades recur with curious persistency throughout the class, as well as others of a sagittate type, in which the basal lobes are pointed rather than rounded (see p. 64). Like the cordate leaf, the "arrowhead" may also be regarded as a modification of the parallel-veined type; it seems to owe its form to a

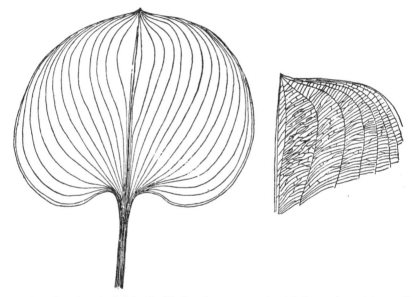

Fig. c. *Eurycles sylvestris*, Salisb.: leaf-limb and upper part of petiole (× 0·25) ; to right, small part of right-hand side of limb near apex (× 0·5). [Arber, A. (1921³).]

tendency on the part of the two main lateral veins to diverge from the mid-rib—a divergence which affects the entire length of the vein, and not merely its basal portion, as in the cordate type. The result is that the apex of each main lateral vein, which in the cordate type returns to the midrib, in the sagittate type points freely away from it, at an angle which often exceeds 120°. An analogous difference in the outline of the limb, due to the convergence or freedom of the veins at the apex, exists between the leaves of certain species of Yam (*Dioscorea*). In the cordate leaf of *D. sativa*, L. (fig. ci, A), for instance, the main veins all approximate at the tip of the leaf, while in *D. quinqueloba*, Thunb. (G) and *D. brachybotrya*, Poepp. (H), the separation of the veins at the apex has rendered possible the formation of a lobed leaf. The way is thus paved for a complete division of the leaflets, as in *D. pentaphylla*, L. (E and F); this leaf sequence in *Dioscorea* thus shows how a compound leaf may be derivable from a parallel-veined type.

The subject of compound leaves leads to the consideration of the controversial case of *Smilax*[1], in which, above the sheath, there is a central stalked blade and two lateral tendrils (fig. cii. 1, p. 128); fig. cii, 3 B shows that

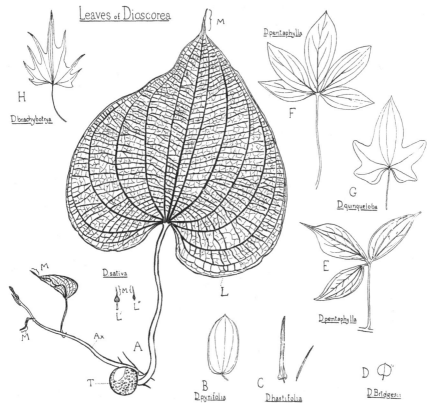

Fig. ci. Leaves of *Dioscorea* (all ×⅓). A, *D. sativa*, L., leaf *L.*, seen from under surface, with axillary shoot *Ax.*; *M.*, mucro, which occupies a large proportion of the small leaves *L'* and *L''*, but a small proportion of the mature leaf, *L.*; the surface of the tuber, *T.*, shows corky flecks. B, *D. pyrifolia*, Kth., from Philippine Islands. C, *D. hastifolia*, Nees, from W. Australia. D, *D. Bridgesii*, Griseb., from Valparaiso. E and F, *D. pentaphylla*, L., E from Kew Gardens, F from India. G, *D. quinqueloba*, Thunb., from Japan. H, *D. brachybotrya*, Poepp. from Chili. (B, C, D, F, G, H, from specimens in Brit. Mus. Herb.) [A.A.]

the tendrils have essentially the same anatomical structure as the median petiole, but, while the tendrils retain their "radial" petiolar structure, the central member, by invagination (fig. 2 B and figs. xcviii, 14 and 15, p. 124), and by expansion, developes its distal region into a blade. It seems to me that this peculiar leaf-form is best explained as due to the origin of three

[1] Arber, A. (1920[2]).

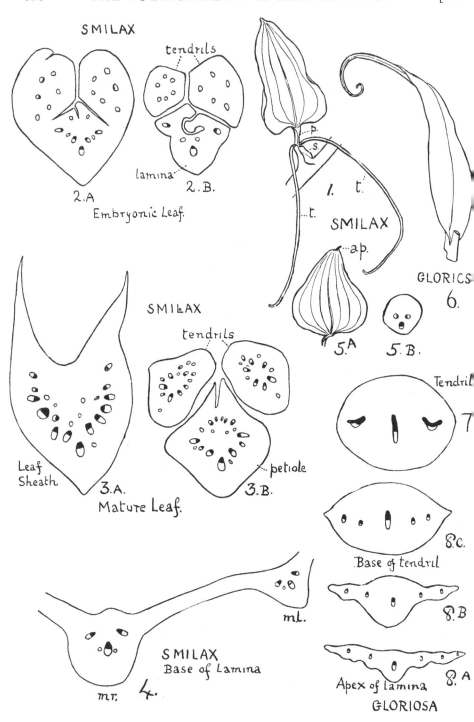

SMILAX

tendrils

lamina

2.A

2.B.

Embryonic Leaf.

p.

s.

1. t.

...t.

SMILAX

...ap.

GLORICS

6.

5.A   5.B.

SMILAX

tendrils

Leaf
Sheath

3.A.

Mature Leaf.

petiole

3.B.

Tendril

7

8.c.

Base of tendril

ml.

8.B

SMILAX
Base of Lamina

8.A

Apex of lamina

mr.   4.

GLORIOSA

Fig. cii. In all sections xylem is indicated in black, and phloem and undifferentiated bundles in white. Fig. 1, *Smilax herbacea*, L.: single leaf showing sheath, *s.*; tendrils, *t.*; and petiole, *p.* (× ⅔). Fig. 2, *Smilax herbacea*, L.: A, transverse section of very young lamina and accompanying tendrils; vascular tissue too young to be well differentiated; petiole as yet undeveloped : B, actual attachment of tendrils; these sections illustrate large size of tendrils relative to lamina in embryonic stage (× 31, *circa*). Fig. 3, *Smilax herbacea*, L.: transverse sections of mature leaf; A, through leaf-sheath; B, through petiole and tendrils, showing relative orientation of organs (× 19, *circa*). Fig. 4, *Smilax herbacea*, L.: transverse section close to base of lamina passing through midrib *mr.*, and main lateral *ml.* (× 19, *circa*). Fig. 5, *Smilax mauritanica*, Poir.: A, pseudo-lamina with thickened apex *ap.* (× ⅔); B, transverse section through apex (× 19, *circa*). Fig. 6, *Gloriosa superba*, L.: leaf with apical tendril (× ⅔). Fig. 7, *Gloriosa superba*, L.: transverse section through tendril; lateral tendril veins fused so that only midrib and two laterals show (× 31, *circa*). Fig. 8, *Gloriosa superba*, L.: transverse sections of another leaf showing transition from apex of limb, A, to base of tendril, C (× 31, *circa*). [Arber, A. (1920²).]

equivalent organs—the two tendrils and the stalked blade—by the basal chorisis of the petiole.

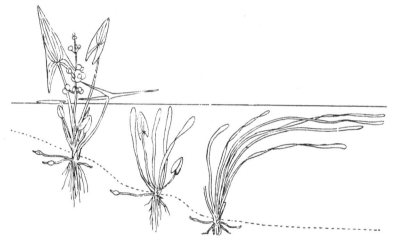

Fig. ciii. Arrowhead (*Sagittaria sagittifolia*, L.). Plants in deep water with ribbon-leaves only, and one in shallow water with leaves mostly aerial and sagittate. Tubers occur on underground shoots. [Lister, G. (1920).]

The curious horn-like projections which in *Tamus communis*, L. (frontispiece) occupy a position corresponding to that of the tendrils in *Smilax*, may possibly be organs homologous with these tendrils, but, as they are non-vascular, it is perhaps more probable that they are emergences. We need, however, more light on these peculiar outgrowths, whose high degree of development and constancy of position seem to call for explanation.

It has been repeatedly observed that a series of transitional forms can be found in *Sagittaria*, connecting the ribbon-like submerged leaves with the mature arrowhead type (fig. ciii). A somewhat similar, but less striking series can be followed in figs. cxlvi, E, F, G, p. 192, drawn from the Aroid *Symplocarpus foetidus*, Nutt. I look upon the intermediate leaves, between the bladeless and bladed types—in these and other Monocotyledons —as affording a picture of the way in which the blade may have been developed from a simple phyllode in the course of evolution; but it must be freely admitted that it is possible to read the series in the opposite direction. It is indeed by no means in every case that we can offer any contribution to a definite proof that the Monocotyledonous blade is a pseudo-lamina; but there are certain considerations which may perhaps be held to point to the general validity of this view.

On looking through long series of herbarium specimens of Monocotyledons, I have gained the impression that there is liable to be a greater range of leaf size—especially as regards width of limb—within a single genus in this class than in Dicotyledons. This point is illustrated for the genus *Dioscorea* in fig. ci, p. 127[1]. Actual measurements can be, of course, of little value in such a connexion, unless made on a large scale and in accordance with statistical methods; but simply as examples to make my meaning clearer, I may mention that the leaves of *Crinum latifolium*, L., may be 11 cms. wide, while those of *C. natans*, R. Br., are under 3 cms. wide; *Alstroemeria nemorosa*, Gardn., again, has leaves up to 3·6 cms. wide, while in *A. apertifolia*, Baker, the corresponding measurement is 0·75 cm. or less. Such a generalisation about the two classes is obviously very difficult, if not impossible, to submit to proof, and I do not wish to lay much stress upon it, since its degree of value is uncertain. But I think it may at least be noted that such variation in leaf width is just what would be expected if one accepts the idea that the "blade" of Monocotyledons arises by greater or lesser degrees of expansion of the petiolar region.

The idea that Monocotyledons are monophyletic can by no means be treated as axiomatic, though the balance of likelihood is perhaps rather in its favour. If it be accepted, however, there seems to me to be a presumption that the original stock, from which the class arose, had leaves which were petiolar phyllodes. In this case we should expect—on Dollo's "Law of Irreversibility"[2]—that the lamina once gone was gone for ever, and it thus becomes probable on general grounds, that any blade-like development, met with in Monocotyledonous leaves of the present day, would be a pseudo-lamina, representing a secondary modification of the surviving petiole or sheath.

---

[1] For figures of other species see Hauman, L. (1916).   [2] See Arber, A. (1919[2]) and (1919[3]).

CHAPTER VI

THE PROPHYLL

(i) THE PROPHYLLS OF NORMAL SHOOTS

BOTH in Dicotyledons and Monocotyledons, the first leaves produced by lateral branches differ so much in character from those which succeed them, as to have received special names—"prophylls," in the case of the vegetative shoots, and "bracteoles," when the shoots are reproductive. They are commonly simple in form, and often seem to be equivalent to leaf-sheaths alone. As a rule Dicotyledons have a pair of prophylls laterally placed, while Monocotyledons have a single prophyll lying between the main axis and the lateral branch, with its back to the main axis, e.g., *Smilax*, figs. civ, 6, 7, p. 132. Exceptions to this rule occur, but are rare. *Aristolochia* and most of the Anonaceae[1] are described as having a single prophyll facing the axillant leaf, so that they resemble Monocotyledons. Figs. 2 and 3 show examples of an approach to the Monocotyledonous condition in *Aristolochia* and the Leguminous genus *Cercis*. *Dioscorea* (figs. 4, A and B), on the other hand, though itself a Monocotyledon, recalls the Dicotyledons in the possession of paired lateral prophylls.

A comparative examination of Monocotyledonous prophylls (in which term I include bracteoles) shows that they very commonly have a two-keeled form, with vascular strands running in the keels, while there is no conspicuous median bundle (figs. civ, 5, 7, 8; cxxxv, 10 A, 15 A, p. 169; cxxxviii, 4 A–C, p. 181). This type of structure has suggested the theory that the prophyll consists of two fused foliar members—a view which has been maintained by various authors. But, though this theory is plausible at first glance, it does not survive a critical study so well as that of the older morphologists who interpreted the prophyll as a single leaf. They regarded the two-keeled form, and the lack of a midrib, as due to the fact that, as the prophyll develops, it is compressed between the main axis on one side and the axillant leaf on the other, and thus finds itself "gênée dans sa croissance[2]." The most telling evidence in support of this view, is, I think, that derived from an examination of the skeletal system of the prophylls. In this connexion special interest attaches to a detailed study of these structures made by Gravis[3] for the case of *Tradescantia virginiana*, L. The prophylls of the buds on the rhizomes of this species are conical, and open by a slit on the side opposite the main axis; they are variable both in size and anatomical complexity.

---

[1] Fries, R. E. (1911).   [2] Duval-Jouve, J. (1877).   [3] Gravis, A. (1898).

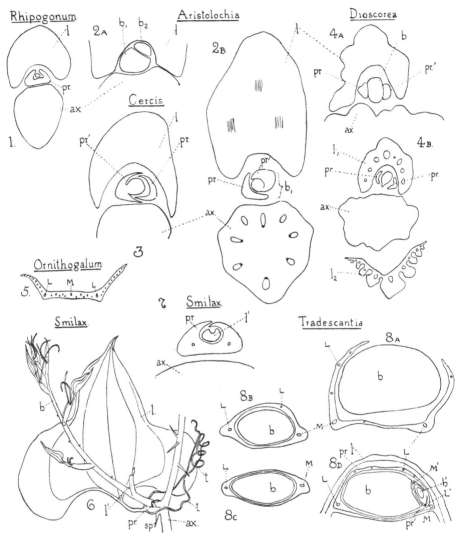

Fig. civ. Prophylls. (Throughout the figures *ax.*, axis; *l.*, leat subtending axillary bud, *b.*; prophylls of bud, *pr.* and *pr.'*; fig. 5 [Lonay, H. (1902)]; fig. 8 [Gravis, A. (1898)]; the remaining figures [A.A.].) Fig. 1, *Rhipogonum album*, R.Br., transverse section through shoot apex (× 23). Figs. 2 A and B, *Aristolochia Clematitis*, L., two transverse sections from series through axil of leaf *l.* (× 47); 2 A, two buds, b_1 and b_2, between *l.* and *ax.*; 2 B, at a slightly higher level, to which the minor bud, b_2, does not reach; the petiole of *l.*, with its bundles, is cut obliquely, owing to the angle at which it leaves the axis. Fig. 3, *Cercis Siliquastrum*, L., transverse section from series through shoot apex (× 23); *pr.* is free from the bud axis, but *pr.'* is still attached. Fig. 4 A, *Dioscorea divaricata*, Blanco, transverse section from series through shoot apex; axillary bud with paired prophylls (× 47); fig. 4 B, *Dioscorea sativa*, L., transverse section near apex of shoot (× 14); one leaf, l_1, cut near its base, subtending axillary bud with paired prophylls; second leaf, l_2, cut at a higher level, shows dorsal invaginations. Fig. 5, *Ornithogalum caudatum*, Ait., prophyll in which main laterals, *L.* and $L_.$, are larger than median bundle, *M.* (× 2). Fig 6, *Smilax aspera*, L., small part of shoot (× ½); the leaf, *l.*, with tendrils, *t.* and *t.*, has branch, *b.*, in its axil; the two-keeled prophyll of *b.* faces *l.*, while *l.'*, the next leaf of *b.*, is superposed to it; *sp.*, spines (April 28, 1921). Fig. 7, *Smilax herbacea*, L., transverse section of bud with prophyll, from series through shoot apex (× 23) to show that the next leaf of the bud, *l.'*, is superposed to the prophyll. Figs. 8 A—D, *Tradescantia virginiana*, L.; fig. 8 A, axillary, *b.*, with prophyll, *pr.*, including seven bundles, *M.*, median, *L.* and $L_.$, the two main laterals (× 5); fig. 8 B, smaller prophyll with three bundles only (× 7½); fig. 8 C, smaller prophyll with two bundles only (× 7½); fig. 8 D, prophyll, *pr.*, of bud, *b.*, in axil of *l.*, with bud, *b.'*, in axil of *pr.*, opposite bundle *M.*; *M.'* and *L.'*, median and lateral bundles of *pr.'*, the prophyll of *b.'* (× 7½).

Frequently there are two nerves, each occupying a slight keel, and the prophyll thus conforms to the type commonest among Monocotyledons (fig. civ, 8 C). Of the two nerves, however, one (M) is a little larger than the other, and is regarded by Gravis as the median bundle, which is displaced by pressure; the smaller bundle (L) he takes to be one of the two main laterals, the second being absent. This interpretation may sound somewhat strained when applied to these prophylls with only two nerves, but it is confirmed by a study of other types, such, for instance, as a three-nerved form in which the midrib and both main laterals are present (fig. 8 B). Still more elaborate prophylls are also to be found, such as a seven-nerved but two-keeled form (fig. 8 A), in which the symmetry can scarcely be explained except on the assumption that the bundles in the keels are respectively the midrib and a main lateral. In addition to these multinerved prophylls, there are reduced and feeble examples with a single strand, which it seems reasonable to interpret as the median bundle.

The prophylls of the Gramineae conform to the *Tradescantia* type. They are commonly two-keeled, and, since the bundle in one of these keels precedes the other in development, and is so placed that it faces the median bundle of the succeeding leaf, it is reasonable to regard it as being, in a morphological sense, the median bundle of the prophyll[1]. Examples of Grass prophylls are to be seen in figs. cv, 19, p. 134; cxxxiv, 8 B, p. 167; cxxxv, 10 A, 13 B, 15 A, p. 169.

In another case which has been studied in some detail—that of *Ornithogalum caudatum*, Ait., described by Lonay[2]—the prophylls, though they are two-keeled, belong to a second type in which there is none of the structural asymmetry which we find in *Tradescantia* and the Grasses; in this second type, it is the two main lateral bundles which occupy the keels (fig. civ, 5). Lonay ascribes the two-keeled form to pressure, and points out that the *foliage-leaves* of the axillary buds, which develop tightly enclosed in the leaf-sheaths, share the two-keeled character, though it is less conspicuous than in the prophylls. In the case of the prophylls, he shows that the median bundle is somewhat smaller than the main laterals. This predominance of the main laterals is not peculiar to prophylls, but is found in various Monocotyledonous foliage-leaves, e.g., those of *Astelia* (fig. lv, p. 77) and a number of Crocoideae (fig. xcv, p. 120)[3]. In addition to the two types of prophyll which we have just distinguished, it is possible that there may be yet a third type, which owes its two-keeled or even bifid form, to the fact that it is a reduced leaf, in which the limb is absent, and nothing is left but a pair of stipules, more or less completely fused. The drawing on p. 94 of a section of an apical bud of *Potamogeton* (fig. lxx, 1 A) shows a ligular

---

[1] Bugnon, P. (1921[1]) ; Arber, A. (1923[3]); Guillaud, M. (1924).   [2] Lonay, H. (1902).
[3] Arber, A. (1921[1]), pp. 328—9.

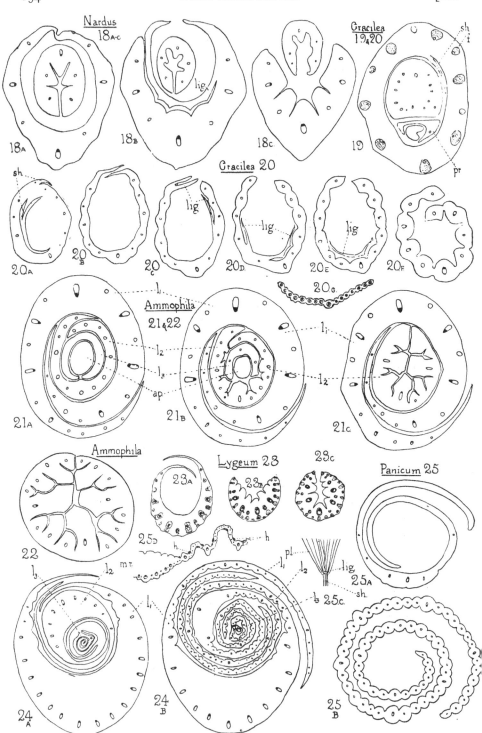

Fig. cv. Grass Leaves. Figs. 18 A—C, *Nardus stricta*, L.: transverse sections through apical bud (× 47); A, closed sheath of outer leaf; B, origin of ligule; C, limb. Figs. 19, 20 A—G, *Gracilea Koyleana*, Hook. f.: fig. 19, transverse section through apical bud (× 77); outer leaf, each of whose bundles is accompanied by mass of fibres, *f.*, is cut through its closed sheath, but one margin, *sh.*, overlaps; *pr.*, prophyll of bud; figs. 20 A—F, series of transverse sections through leaf from bud (× 77), passing from attachment to axis, A, through development and detachment of ligule, C—E, to limb, F; fig. 20 G, transverse section of limb of mature leaf (× 14). Figs. 21 A—C and fig. 22, *Ammophila arundinacea*, Host.: fig. 21 A—C, transverse sections from series through apical bud from Wells, Norfolk, to show origin of invaginations in leaf, l_2; fig. 22, transverse section of young leaf from another bud (× 47). Figs. 23 A—C, *Lygeum Spartum*, Loefl.: transverse section through sheath, A, base of limb, B, and higher in limb, C, of mature leaf (× 14). Figs. 24 A, B, *Saccharum officinarum*, L.: transverse sections through three leaves l_1, l_2, l_3, of apical bud, to show appearance of invaginations (× 23). Figs. 25 A—D, *Panicum plicatum*, Lam.: A and B, transverse sections of sheath and limb of developing leaf (× 47); C, junction of sheath, *sh.*, and plicate limb, *pl.*, of mature leaf, showing ligule, *lig.* (× ½); D, transverse section close to base of mature limb, showing part of thickened midrib, *mr.*, and of one side of limb; *h.*, hinge cells (× 14). [Arber, A. (1923³).]

sheath (*lig. s_1*) which is definitely two-keeled, with a conspicuous wing (*k.* and *k.'*) at each keel. The ligular sheath of *Laxmannia grandiflora*, Lindl. (fig. lii, 19 B, p. 74) is of a similar type, as is also the axillary stipule of the Dicotyledon, *Victoria regia*, Lindl.[1]. Such stipular or ligular sheaths develop in a confined space between the leaf to which they belong and the axis, and thus—like prophylls—they probably owe the peculiarities of their form to the pressure to which they must submit during ontogeny. They afford a proof that an organ which is merely a part of a single leaf, may yet be as obviously two-keeled as any of the prophylls whose binerved structure has been claimed as evidence of duality.

If the Monocotyledonous prophyll were the equivalent of two leaves whose midribs formed the keels, we should expect to find that it would reveal its bifoliar nature by subtending two axillary buds. This, however, proves not to be the case. It appears, especially from the work of Rüter[2], that it is usual for the prophylls to possess one axillary bud only, and that this bud is nearly always so placed as to be opposite one of the keels. In *Tradescantia*, in which one keel is occupied by the median strand and the other by one main lateral, there is, according to Gravis, a tendency for the axillary bud to arise in the morphologically normal position, i.e. opposite the strand which he interprets as median (*b.'* in fig. civ, 8 D, p. 132), and among the Grasses, the same relation appears to be the rule. Fig. cxxxv, 10 A, p. 169, shows a lateral bud of the second order, b_2, in the axil of a prophyllar leaf, *pr.*, of the Wheat, *Triticum vulgare*, Vill.; the bud arises opposite that bundle of the prophyll which, on other grounds, one would suppose to be morphologically median.

[1] Glück, H. (1919).   [2] Rüter, E. (1918).

(ii) The Prophylls of "Short Shoots"

(a) The Phylloclades of Myrsiphyllum

Asparagus medeoloides, Thunb. (*Myrsiphyllum asparagoides*, Willd.) is a climbing plant from South Africa, whose green trailing shoots (fig. cvi, I A) are

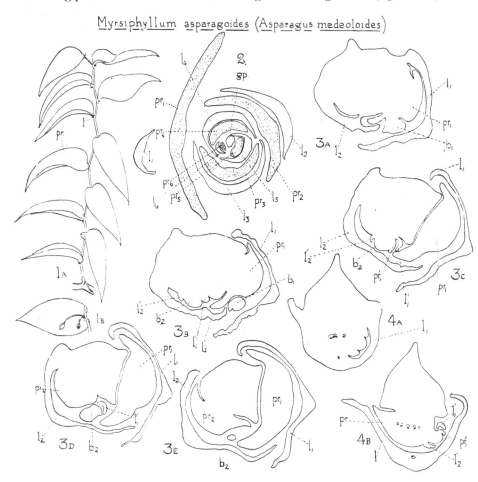

Myrsiphyllum asparagoides (Asparagus medeoloides)

Fig. cvi. *Myrsiphyllum asparagoides*, Willd. (*Asparagus medeoloides*, Thunb.). Fig. I A, vegetative shoot, La Mortola, October (reduced); Fig. I B, phyllode with inflorescence below, February (reduced). Fig. 2, transverse section through plumule apex ($\times 47$); *g.p.*, undifferentiated apical cone; pr_1—pr_6, phylloclades (dotted) in axils of leaves l_1—l_6 (white). Figs. 3 A—E, series of transverse sections through the tip of plumule ($\times 47$), showing two leaves, l_1 and l_2, with their phylloclades, pr_1 and pr_2, and buds, b_1 and b_2; b_2 bears leaves l_1', l_2' and phylloclade pr_1'. Figs. 4 A and B, two sections from series showing detachment of a bud with its phylloclade, *pr.*, and leaves l_1' and l_2'; l_1' bears phylloclade pr_1' in its axil. [A.A.]

sold by florists under the mistaken name of "Smilax". All botanists, hitherto, seem to have treated the assimilating organs (*pr.*, fig. cvi, I A)—by analogy

with the "needles" of *Asparagus*[1]—as stem structures, but an examination
of them, and especially the study of their relations as seen in serial sections
through young shoots, has convinced me that these organs are, in reality,
true leaves[2]; I interpret them as the prophylls of "short shoots" arising in
the axils of scale leaves, *l*. This view may seem difficult to accept for such a
branch as that shown in fig. cvi, I A, since here the axillary shoots bear prophylls
only, and do not display, in addition, even the least rudiment of a growing
point. But there are other cases, which seem to me to confirm my interpretation; in these the growing point does not become abortive, but carries on
the development, so that, instead of a reduced "short shoot", a normal, long,
axillary shoot is produced, bearing, at its extreme base, a phylloclade as its
first foliar organ. Branches of this type are shown in the sections drawn in
figs. cvi, 3 and 4. In figs. 3 A—E, the shoot b_1, arising in the axil of the
leaf l_1, bears the prophyll, pr_1, after which its growth is continued for a short
distance, while the shoot, b_2, arising in the axil of the leaf l_2, bears the
prophyll, pr_2, and also a succession of rudimentary leaves, of which l_1' has
detached itself in fig. 3 D. Such a shoot is seen in fig. 4 B in a more advanced
state, in which it bears not only the prophyll, pr., but the leaves l_1', and l_2',
and a rudiment, pr_1', which is probably the phylloclade axillary to l_1'. These
sections are puzzling at first glance, partly because of the exaggerated size
of the prophyll, which forces the growing point to one side, and partly because the back of the prophyll remains fused with the main axis after the
rest of the shoot, to which it belongs, is free. As will be seen in fig. cvi, 4 B,
the xylem of the vascular bundles of the phylloclade faces *towards the axillant leaf.* That is to say, the strands occupy the position which one would
naturally expect on the assumption that the phylloclade is the prophyll of
the axillary shoot; on any other theory, however, this orientation becomes
difficult to understand.

(b) The Phylloclades of the Rusceae[3]

The tribe Rusceae (Liliaceae) consists of three genera, *Danae, Ruscus* and
Semele. Like the Asparageae, to which they are related, these genera
display a peculiar vegetative morphology. *Danae racemosa,* (L.) Mönch., the
Alexandrian or Victor's Laurel, which is grown in this country for the sake
of its singularly graceful evergreen shoots, has assimilating organs (phylloclades), shown in figs. cvii, I A, p. 138, and cviii, 2 A, p. 139, which closely
resemble those of *Myrsiphyllum asparagoides.* They arise in the axils of
scale-leaves and are so orientated that the xylem groups of the bundles are
directed towards the axillant leaf (figs. cviii, 2 C, 4 B, 5, 6 A). In *Danae* the
inflorescence is independent of the phylloclade and takes the form of a
terminal raceme (fig. cviii, 3 A).

[1] See pp. 51—53.   [2] Arber, A. (1924[3]).
[3] For a fuller discussion of this subject with references, see Arber, A. (1924[1]).

A more familiar member of the tribe is our British *Ruscus aculeatus*, L.—
the Butcher's-broom or Knee-holly. The stiff prickly "foliage" of this plant

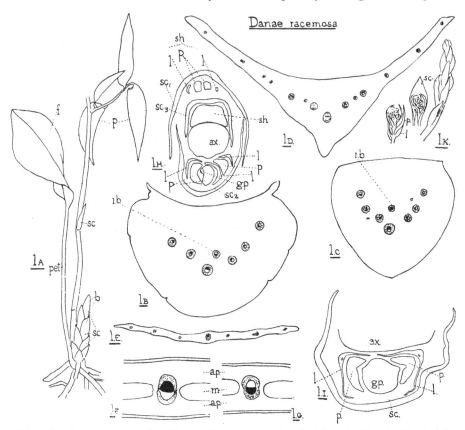

Fig. cvii. *Danae racemosa*, (L.) Mönch. Fig. 1 A, part of a plant from La Mortola, Ventimiglia,
March (× ½); *b.*, winter bud; *sc.*, scale leaves; *f.*, primary foliage leaf with long petiole, *pet.*; *p.*,
phylloclades; figs. 1 B—E, transverse sections through leaf *f.* in fig. 1 A; *i.b.*, bundle with xylem
turned downwards (× 14); fig. 1 B, top of short basal sheath; fig. 1 C, high in petiole; fig. 1 D,
base of limb; fig. 1 E, near apex of limb; fig. 1 F, small part of fig. 1 E (or similar section) to show
single vein of foliage leaf (× 77) for comparison with fig. 1 G, which shows, on same scale, a single
vein from near apex of one of the phylloclades, *p.*, in fig. 1 A; *a.p.*, assimilating parenchyma; *m.*,
colourless large-celled mesophyll; fig. 1 H, transverse section from microtome series through central
part of bud, such as *b.* in fig 1 A (× 23), including axis, *ax.*; three scale leaves, *sc₁*, *sc₂*, *sc₃*,
each subtending a bud. The bud, *sh.*, inside *sc₃* is cut at a level below its appendages; the buds
inside *sc₁* and *sc₂* show leaves, *l.*, with phylloclades, *p.*, in their axils; *g.p.*, growing apex; fig. 1
I, bud in axil of scale leaf, *sc.*, similar to the three buds shown in fig. 1 H, but rather older. The
growing axis of the bud, *g.p.*, has produced a pair of prophyllar leaves, *l.*, subtending phylloclades,
p.; fig. 1 K, tip of bud, such as *b.* in fig. 1 A, later in the spring (reduced); to the left, one of the
axillary branches of this shoot with its terminal inflorescence is drawn from the adaxial side, with
and without its axillant scale-leaf (× 1). [Arber, A. (1924¹).]

has of late found its way into commerce: after its beauty has been effectually destroyed by aniline dyes, it is sold for winter decoration. As in
Danae, sections of the phylloclades show that the xylem of the bundles is

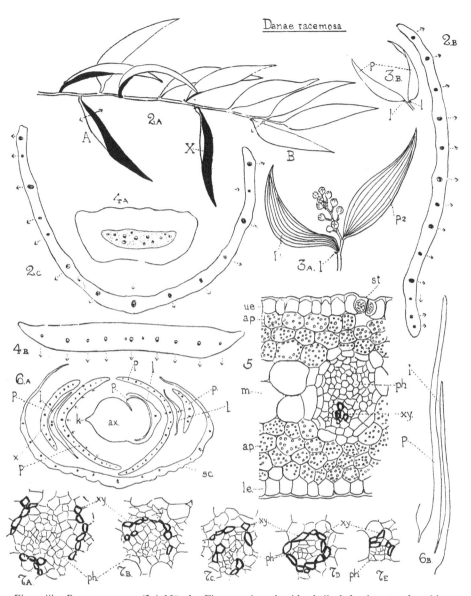

Danae racemosa

Fig. cviii. _Danae racemosa_, (L.) Mönch. Fig. 2 A, branch with phylloclades in natural position, Oct. 4 ($\times \frac{1}{2}$); the surface of each phylloclade towards its axillant leaf is left white, and surface away from axillant leaf marked black; fig. 2 B, transverse section, at level of arrow, of phylloclade B, which is held edgeways ($\times 14$); the dotted arrows indicate the direction in which xylem of bundles faces; fig. 2 C, transverse section, at level of arrow, of phylloclade A, which is held in unusual position, with surface away from axillant leaf uppermost ($\times 14$). Fig. 3 A, shoot bearing leaf _l._, in whose axil phylloclade p_1 arises; a second phylloclade, p_2, whose axillant leaf is invisible; a terminal raceme of young fruits, still enclosed in persistent perianths ($\times \frac{1}{2}$); fig. 3 B, end of sterile shoot, with two phylloclades, _p._, in axils of leaves, _l._; no terminal phylloclade ($\times \frac{1}{2}$). Figs. 4 A and B, transverse sections of a phylloclade ($\times \frac{1}{2}$); fig. 4 A, stalk; fig. 4 B, near base of expanded region. Fig. 5, small segment of transverse section of a young phylloclade ($\times 318$); fibres of bundle-sheath still thin-walled; _xy._, xylem; _ph._, phloem; _st._, stomates; _u.e._, upper epidermis; _l.e._, lower epidermis; _a.p._, assimilating parenchyma; _m._, colourless mesophyll. Fig. 6 A, transverse section of young bud (May 24) in axil of scale-leaf, _sc._; its axis, _ax._, bears two rows of leaves, _l._, subtending phylloclades, _p._ The leaves in whose axils the two outermost phylloclades arise, do not reach to this level. The keel, _k._, of the axis corresponds to the midrib of the next leaf above ($\times 14$). Fig. 6 B, radial longitudinal section through leaf _l._ with phylloclade _p._ in its axil, from microtome series through bud, May 24 ($\times 14$). Figs. 7 A—E, series of transverse sections ($\times 318$) to show history of the single bundle of the leaf in whose axil the phylloclade marked with a $\times$ in fig. 6 A arises; fig. 7 A, near leaf base; fig. 7 E, near leaf apex. [Arber, A. (1924[1]).]

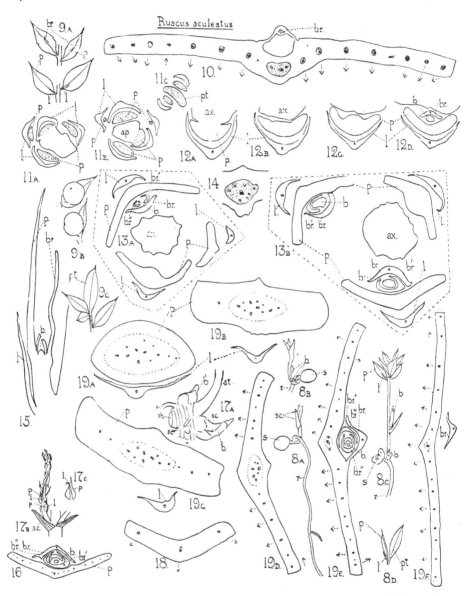

Fig. cix. *Ruscus aculeatus*, L. Figs. 8 A—D, seedlings grown by Mr T. A. Dymes, Feb. 21; figs. 8 A and B, young subterranean seedling, in which only scale-leaves, *sc.*, have developed, viewed from the two sides (× ½); in fig. 8 B, three lateral buds, *b.*, are visible; *s.*, seed; figs. 8 C and D, older seedling bearing phylloclades, *p.*, in axils of scale-leaves, *l.*; two lateral buds, *b.*; seed shrunken as compared with figs. 8 A and B (× ½); fig. 8 D, two uppermost phylloclades of 8 C, one in axil of leaf, *l.*, while the other, *p.t.*, is terminal (× 1). Fig. 9 A, two successive phylloclades, *p.*, of mature shoot viewed from the two sides; *l.*, axillant leaves; *br.*, bracts borne on phylloclades (× ½): fig. 9 B, two views of a phylloclade bearing a fruit, Feb. (× ½); fig. 9 C, end of shoot showing terminal phylloclade, *p.t.* (× ½). Fig. 10, transverse section of phylloclade at level of its junction with bract (× 14). Arrows show directions in which xylem groups of bundles point; bundle marked with a × is amphivasal; axillant leaf would lie to foot of page. Figs. 11 A—C, transverse sections from series through lateral bud, May 13 (× 23); *l.*, leaves (white); *p.*, phylloclades (dotted); fig. 11 C shows farther development of *ap.*, growing apex, in 11 B, with the formation of two leaves subtending phylloclades, and a third phylloclade which is terminal, *p.t.* Figs. 12 A—D, series of sections through leaf, *l.*, with phylloclade, *p.*, in its axil, May 13 (× 23); fig. 12 A shows phylloclade not yet detached from axis, *ax.*, while fig. 12 D shows bract, *br.*, and its axillary bud, *b.*, on side of phyllo- clade remote from axillant leaf. Figs. 13 A and B (enclosed in dotted lines), transverse sections through buds, June 2 (× 14), in each case showing axis, *ax.*, and three leaves, *l.*, with phylloclades, *p.*, in their axils. One phylloclade in each figure is cut just above the attachment of the bract, *br.*, which arises on its adaxial surface, and subtends bud, *b.* The bud in fig. 13 A has two first leaves, *br.′* and *br.″*, which are lateral and opposite, while those in fig. 13 B, on the other hand, have one larger, obliquely-placed first leaf, *br.′.* Fig. 14, transverse section of midrib of a sterile phylloclade (× 23). Fig. 15, radial longitudinal section of leaf, *l.*, with phylloclade, *p.*, in its axil, bearing bract, *br.*, and bud, *b.*; June 2 (× 14). Fig. 16, transverse section of phylloclade, June 16, bearing bract, *br ,* in whose axil is the bud, *b.*, bearing leaves, *br.′*, and *br.″*, arranged as in the buds in fig. 13 B (× 14). Fig. 17 A, end of rhizome, *rh.*, marked with scars of former scale-leaves, *sc.′*, and bearing roots, *r* , and terminating in aerial stem, *st.* The two winter buds, *b.* and *b.′*, clothed with scales, *sc.*, will develop into the shoots of the current year; Feb. 6, 1923 (× ½); figs. 17 B and C, part of young shoot later in the spring, probably May; fig. 17 B, branch in axil of scale-leaf, *sc.*, bearing leaves, *l.*, and phylloclades, *p.* (× ½); fig. 17 C, leaf and phylloclade of fig. 17 B (× 1). Fig. 18, transverse section near apex of a young phylloclade (June 16) to show reduction of bundles to three; axillant leaf would lie towards lower edge of page (× 23). Figs. 19 A—F, series of transverse sections from base upwards through leaf, *l.*, and its phylloclade, *p.*, June 16 (× 14); figs. 19 C and D show the process of torsion that brings the phylloclade into the vertical position in figs. 19 E and F. The arrows show direction of xylem in bundles; fig. 19 E passes through the attachment of bract, *br.*, in whose axil the bud, *b.*, arises, bearing leaves, *br.′*, *br.″*, *br.‴*; fig. 19 F is above attachment of bract. [Arber, A. (1924[1]).]

directed towards the axillant leaf. The phylloclades diverge even more from normal leaves than do those of *Danae*, for the inflorescence is produced from their surface in the median line (fig. cix, 9 B). The buds which will give rise to inflorescences are seen in section in many of the sketches in fig. cix.

The other species of *Ruscus*, *R. Hypoglossum*, L. (fig. cx, p. 142) and *R. Hypophyllum*, L. (fig. cxi, p. 143) are similar to *R. aculeatus* in the general features of their structure. Here again the inflorescence is borne on the phylloclade in the median line, either on the adaxial or abaxial surface.

The monotypic *Semele androgyna*, (L.) Kunth, differs from *Ruscus* in being a climbing shrub. The long *Asparagus*-like Spring shoots (fig. cxii, 30 A, p. 144) give rise to lateral branches, each of which bears two series of scale leaves with phylloclades in their axils (fig. cxii, 31 B). The phylloclades, as a rule, produce several marginal groups of flowers (fig. cxiii, 42, p. 145), but the inflorescence may, on the other hand, be borne on either surface in the median line, as in *Ruscus* (fig. cxii, *X* in 33 G, and *Y* in 33 H, and fig. cxiii, 44 B). When the veins of the phylloclade contain one bundle only, that bundle has its xylem directed towards the axillant leaf (fig. cxiii, 35), but the veins are often, as in *Ruscus Hypoglossum* and *R. Hypophyllum*, supplied by radial groups of strands, instead of single bundles (fig. cxiv, 45 B, p. 146).

The question of the interpretation of the phylloclades of the Rusceae has been the subject of much controversy. The simplest case seems to be that of *Danae*. Here the resemblance to *Myrsiphyllum* is so close, that, if the pro- phyll-interpretation, outlined on pp. 136—137, be accepted for the latter genus, it must necessarily apply also to *Danae*. The foliar view of the phyl- loclades of *Danae* is, moreover, supported by the resemblance of these

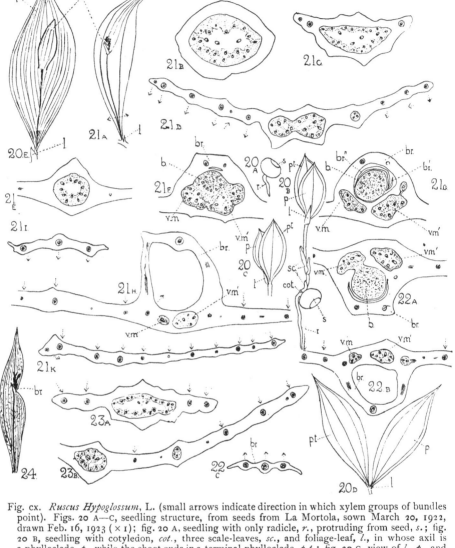

Fig. cx. *Ruscus Hypoglossum*, L. (small arrows indicate direction in which xylem groups of bundles point). Figs. 20 A—C, seedling structure, from seeds from La Mortola, sown March 20, 1922, drawn Feb. 16, 1923 (× 1); fig. 20 A, seedling with only radicle, *r.*, protruding from seed, *s.*; fig. 20 B, seedling with cotyledon, *cot.*, three scale-leaves, *sc.*, and foliage-leaf, *l.*, in whose axil is a phylloclade, *p.*, while the shoot ends in a terminal phylloclade, *p.t.*; fig. 20 C, view of *l.*, *p.*, and *p.t.* from the opposite side; fig. 20 D, apical part of mature shoot (× ⅓) to show terminal phyllo-clade, *p.t.*, standing apparently opposite to a phylloclade, *p.*, in axil of leaf, *l.*; fig. 20 E, phyllo-clade, *p.*, bearing leathery bract, *br.*, on side away from axillant leaf, *l.*; Brit. Mus. Herb. (probably var. *macroglossa*) (× ½). Fig. 21 A, phylloclade (venation shown in less detail than in fig. 20 E) bearing bract, *br.*, and remains of inflorescence on surface away from axillant leaf, *l.* (× ½); figs. 21 B, C, D, E, H, I, K, anatomical structure of phylloclade drawn in fig. 21 A; figs. 21 F and G, sup-plementary stages from another phylloclade (all × 14); fig. 21 B, near extreme base of phylloclade; fig. 21 D, at level of arrow in fig. 21 A; fig. 21 C, intermediate; fig. 21 E, midrib region higher up; ridge above vascular cylinder is first indication of bract; fig. 21 F, section from another phylloclade at a slightly higher level, showing bract, *br.*; vascular system, *b.*, for bud axis; *v.m.* and *v.m.'*, vascular masses destined for main veins of upper part of phylloclade; fig. 21 G, same phylloclade as in fig. 21 F, cut at a slightly higher level; bud now free, bearing two leaves, *br.'* and *br."*; fig. 21 H, section from phylloclade in fig. 21 A, cut near base of leathery bract; bud has fallen out; fig. 21 I, bract after detachment; fig. 21 K, section near apex of phylloclade. Figs. 22 A—C, suc-cessive sections (× 14) through bract-bearing region of another phylloclade, in which the inflorescence was on the side towards the axillant leaf; these sections correspond to figs. 21 G, H, I. Figs. 23 A and B, transverse sections of a sterile phylloclade, near the base and a little higher (× 14). Fig. 24, an abnormal phylloclade (× ½), drawn from side away from axillant leaf. (Figs. 21—24, material from La Mortola, March.) [Arber, A. (1924[1]).]

organs to the foliage-leaves of related Liliaceae, such as those sketched in figs. cxv A—D, p. 147, and still more strikingly by comparison with the undoubted foliage-leaves, which *Danae* itself sometimes produces from its

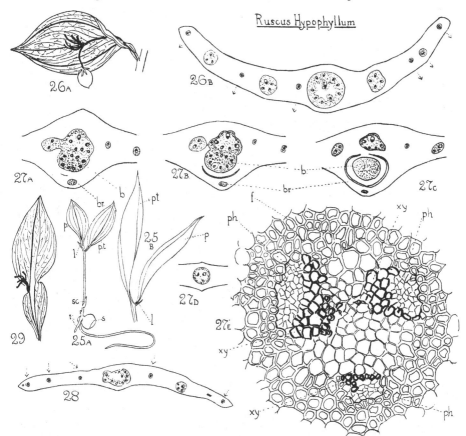

Fig. cxi. *Ruscus Hypophyllum*, L. (material from La Mortola). Fig 25 A, seedling (× ½); seed from La Mortola, sown March 20, 1922, drawn Feb. 16, 1923; *s.*, seed; *r.*, roots; *l.*, leaf with phylloclade, *p.*, in its axil; *p.t.*, terminal phylloclade. The main root, which should be vertical, bent to one side to economise space. Fig. 25 B, end of a mature shoot, to show terminal phylloclade, *p.t.* (× ½). Fig. 26 A, phylloclade with remains of inflorescence and one fruit on side towards axillant leaf (× ½); fig. 26 B, transverse section of a fertile phylloclade near the base (× 14). Figs. 27 A—C, series of three sections of a phylloclade showing origin of bract, *br.*, and vascular cylinder of bud, *b.* (× 14); figs. 27 D and E, transverse section of midrib of phylloclade above inflorescence (× 14 and 193); *xy.*, xylem; *ph.*, phloem; *f.*, fibres. Fig. 28, transverse section of sterile phylloclade not far from base (× 14). Fig. 29, abnormal phylloclade with inflorescence at margin, flowers fallen (× ½). [Arber, A. (1924¹).]

rhizome (fig. cvii, I A, *f.*, p. 138). If such a leaf is represented diagrammatically by figs. cxv, G and H, we may consider the scale-leaves, which *Danae* frequently bears (fig. cxv, E), as derived from this type by loss of the distal region (petiole and blade), while the phylloclade (fig. cxv, I) is formed by a converse mode of reduction—the loss of the proximal region (sheath).

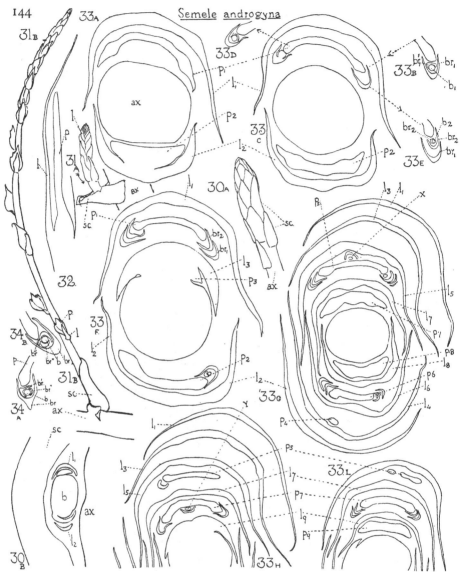

Fig. cxii. *Semele androgyna*, (L.) Kunth. (Material from Cambridge Botanic Garden.) Fig. 30 A, apex of current year's shoot, Jan. 26 ($\times \frac{1}{2}$); *sc.*, scale-leaves; fig. 30 B, transverse section through bud, *b.*, in axil of one of the scale-leaves, *sc.*, from shoot similar to that drawn in fig. 30 A; Feb. 6 ($\times 23$). Figs. 31 A and B, the result of farther development of a bud such as *b.* in fig. 30 B; fig. 31 A, Feb. 19; fig. 31 B, May 4; the scale-leaves, *sc.*, correspond in figs. 30 A and B, and 31 A and B. In fig. 31 B the two series of leaves, *l.*, are seen to bear phylloclades, *p.*, in their axils ($\times \frac{1}{2}$). Fig. 32, a phylloclade, *p.*, and its axillant leaf, *l.*, in radial longitudinal section from microtome series through apex of a shoot similar to that shown in fig. 31 B; May 4 ($\times 14$); the bracts borne on the phylloclade do not lie in the plane of the section. Figs. 33 A—I, transverse sections from microtome series through apex of shoot similar to that drawn in fig. 31 B, May 4 ($\times 14$) showing the development, in two opposite rows, of leaves, l_1—l_9, with phylloclades, p_1—p_9, in their axils; figs. 33 B, C, E, show the development of bracts, br_1 and br_2, axillary buds, b_1 and b_2, and scale-leaves, br'_1 and br'_2, from right-hand margin of p_1, and figs. 33 C and D, from left-hand margin of p_1. In fig. 33 G, p_5 is abnormal in having a bract and bud X in the median region, on the surface towards the axillant leaf; its apex, fig. 33 I, is bifurcated. In fig. 33 H, p_7 also has a bract and bud Y in its median region, but on the surface remote from axillant leaf; p_7 does not bifurcate at apex. In figs. 33 H—I, only part of section shown. Figs. 34 A and B, transverse sections of margins of two phylloclades in which axillary bud is more advanced than in fig. 33, and bears two scale-leaves, *br.'* and *br."* ($\times 14$); fig. 34 A, from material gathered May 5; fig. 34 B, from upper phylloclade shown in fig. cxiii, 35, but at a lower level; June 2. [Arber, A. (1924[1]).]

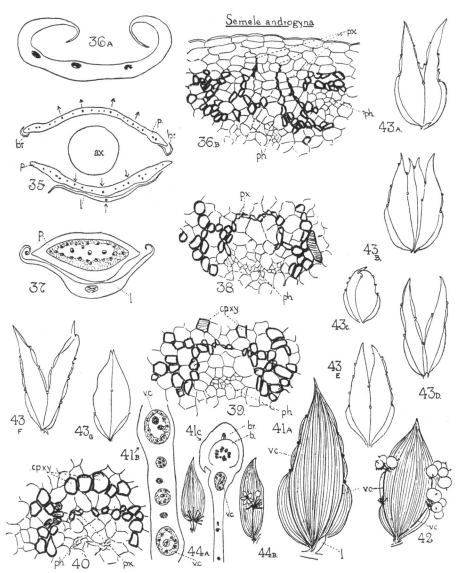

Fig. cxiii. *Semele androgyna*, (L.) Kunth. Fig. 35, transverse section through axis, *ax.*, of bud, June 2, slightly older than that shown in fig. cxii, 33, bearing two phylloclades, *p.*, the axillant leaf, *l.*, of the younger one being alone visible at this level; the arrows show direction of xylem in leaf and phylloclades (× 14). Fig. 36 A, transverse section (× 14) of axillant leaf, *l.*, shown in fig. 41 A (bottom of page); fig. 36 B, median bundle from fig. 36 A (× 193). Fig. 37, transverse section of base of a mature phylloclade, *p.*, with axillant leaf, *l.* (× 14). Figs. 38, 39, 40, bundles from three young axillant leaves, June 2, in which a median strand alone is developed; *px.*, protoxylem; *ph.*, phloem; *cp. xy.*, centripetal xylem (× 318). Fig. 41 A, phylloclade bearing buds, only longitudinal veins indicated (× rather more than $\frac{1}{2}$); fig. 41 B, section of median region of a phylloclade similar to 41 A, not far from base, to show main lateral veins, *v.c.*, which will supply inflorescence (× 14); fig. 41 C, transverse section of margin of phylloclade drawn in fig. 41 A, at base of one of marginal buds, *b.*, showing bract, *br.*, and vein, *v.c.*, from which vascular system of bud is derived (× 14). Fig. 42, phylloclade bearing fruits (× rather more than $\frac{1}{3}$), Feb. Figs. 43 A—G, abnormal phylloclades, all from one plant in Camb. Bot. Garden, Jan. 1923 (× $\frac{1}{2}$); buds indicated by small circles; in A, D, F, the surface towards axillant leaf is shown; in B, C, E, G, the surface away. Figs. 44 A and B, sketches from herbarium material of two phylloclades bearing flowers on the upper surface instead of at margin (× $\frac{1}{2}$). [Arber, A. (1924[1]).]

The interpretation of the phylloclades of *Ruscus* and *Semele* presents more difficulty than the case of *Danae*. The fact that in these genera the inflorescence is borne upon the phylloclade, has led the great majority of

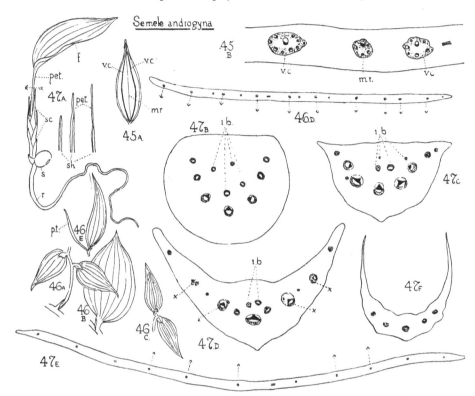

Fig. cxiv. *Semele androgyna*, (L.) Kunth. Fig. 45 A, small but strongly-developed sterile phylloclade, from side towards axillant leaf, to show pair of principal veins, *v.c.*, and less prominent midrib, *m.r.* (× ½); fig. 45 B, transverse section of median region near base of a sterile phylloclade similar to fig. 45 A (× 23). Figs. 46 A—C, sterile phylloclades from feebly-developed basal shoot; A, from side towards axillant leaf; B, from side away from axillant leaf to show well-marked stalk; C, to show termination of shoot (× ½); fig. 46 D, transverse section near apex of feebly-developed sterile phylloclade, similar to those drawn in figs. 46 A—C (× 14); fig. 46 E, end of a shoot from the upper part of the plant (not a feeble shoot from the base such as that used in figs. 46 A—D) to show terminal phylloclade, *p.t.* (× ½). Fig. 47 A, seedling (× ½), Kew, Feb. 7, 1923; seed, *s.*; radicle, *r.*; scale-leaves, *sc.*; first foliage-leaf, *f.*, with petiole, *pet.* Some of the scale-leaves drawn separately (to the right) to show transition between those with sheath, *sh.*, alone, and those with sheath and petiole. Figs. 47 B—D, transverse sections through petiole and base of limb of leaf *f.* in fig. 47 A (× 23); fig. 47 B, low in petiole; fig. 47 C, at level of arrow in petiole; fig. 47 D, at base of limb. The bundles marked *i.b.* have their xylems directed downwards; the veins in fig. 47 D marked with a × contain more than one bundle and are more or less radial. Fig. 47 E, transverse section near apex of same leaf (× 14); fig. 47 F, transverse section of sheath region of one of transitional scale-leaves in fig. 47 A, slightly reconstructed at margins, as section imperfect (× 23). [Arber, A. (1924[1]).]

botanists to treat it as a flattened axis, but I believe that the best interpretation is that first put forward long ago by Duval-Jouve[1] and van Tieghem[2],

[1] Duval-Jouve, J. (1877).   [2] Tieghem, P. van (1884).

but which has been somewhat neglected in recent years. According to the modification of this view which I should wish to adopt, the phylloclade consists of a lateral shoot completely adnate to its own prophyll; but the prophyll predominates in the partnership to such an extent that the anatomy remains entirely foliar. We have in *Tilia* an analogy for such intimate fusion between bracteole and inflorescence-axis (fig. cxvi, p. 148). In the Lime-tree, as in the Rusceae, the anatomy affords no evidence for the adnation; the bracteole is master of the situation, and the structure, below as well as above the exit of the peduncle, is foliar rather than axial.

The chief difficulty in the way of accepting the prophyllar view of the

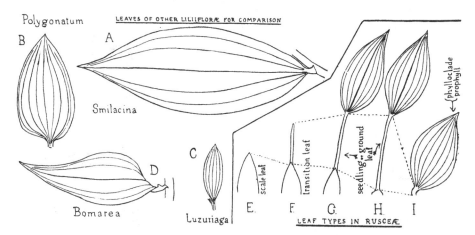

Fig. cxv. A—D, leaves for comparison with phylloclades. A, *Smilacina racemosa*, Desf. (× ½); B, *Polygonatum officinale*, All. (× ½); C, *Luzuriaga* sp. (× 1); D, *Bomarea* (garden hybrid) (× ½). E—I, diagram of different leaf-types met with in the Rusceae, to illustrate interpretation of phylloclade here put forward; E, scale-leaf, consisting of sheathing leaf-base only; F, transitional leaf (e.g. seedling of *Semele*); G, foliage-leaf with well-developed sheath, petiole and limb (such leaves are borne by seedling and sometimes by rhizome of *Danae* and *Semele*); H, similar leaf with reduced sheath (e.g. the leaf figured here for seedling of *Semele*, fig. cxiv, 47 A, p. 146); I, phylloclade (e.g. those of *Danae*, and sterile phylloclade of *Semele*, fig. cxiv, 46 B, p. 146). The dotted lines show the limit of sheath, stalk and limb in the different leaf-categories. [Arber, A. (1924[1]).]

phylloclade of the Rusceae, is the fact that the flower-buds are, in certain species, developed on the adaxial surface, or indifferently on either side of the phylloclade. I am inclined, however, to regard this as a symptom of the extreme completeness with which the identity of the lateral axis has become merged into that of its own prophyll.

The history of the different views which have been held about the phylloclades of the Rusceae has a certain psychological interest. In early days everyone took them at their face value, and treated them as leaves. Then, when morphology began to lay down a definite scheme of organography, the phylloclades were transferred to the category of caulomes, in order to bring them into line with the law that *any organ arising in the axil of a leaf must be an axis.* But, as van Tieghem[1] pointed out, this generalisation does not, as it stands, justify itself, and it becomes necessary to substitute the word *shoot* for *axis.* Now, in the shoot, the axis may be insignificant, while the leaves are relatively important, and when this condition is carried to its extreme limit, we get a case such as that of the Rusceae, in which the activity of the axillary bud may be confined to the production of one leaf —the prophyll. So it becomes clear that the canons of morphology, rightly understood, are not violated by the interpretation of the phylloclade as a

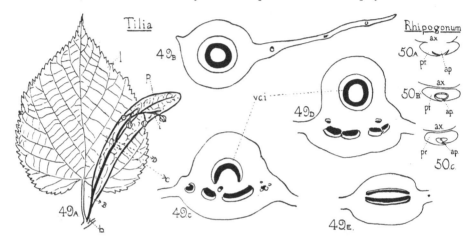

Fig. cxvi. Prophylls for comparison with phylloclades of Rusceae. Figs. 49 A—E, *Tilia* sp.; fig. 49 A, leaf, *l.*, with axillary inflorescence, *i.*, partially fused with its bracteole, *p.*; *b.*, bud in axil of second bracteole (× ½); figs. 49 B—E, series of transverse sections (× 23) showing diagrammatically the plan of the vascular structure of midrib of bracteole ; xylem, black; phloem, white : fig. 49 B, at level of arrow *B*; fig. 49 C, at level of arrow *C*; fig. 49 D, at level corresponding to arrow *D*, but from another bracteole ; the vascular tissue marked *v.c.i.* in C and D is destined for the inflorescence; fig. 49 E, at a level between D and apex of bracteole, but from another bracteole. Figs. 50 A—C, *Rhipogonum album*, R.Br., sections from series from below upwards through very young axillary bud. Axillant leaf would lie towards lower margin of page; *ax.*, axis; *pr.*, prophyll; *ap.*, growing apex of bud (× 23). [Arber, A. (1924[1]).]

prophyll, and we are thus able to return again to the view expressed by Theophrastus[2], more than two thousand years ago, and to regard the phylloclade as being, in reality, what it obviously is in appearance—a leaf.

(c)   The "Fronds" of the Lemnaceae

The flat green assimilating organs of the Duckweeds—*Spirodela, Lemna* and *Wolffia*—have been very variously interpreted, some authors treating them as wholly axial, others as wholly foliar, while some, again, regard them as axial at the base and foliar in the distal region. The view which I hold, and for which I have given detailed reasons elsewhere[3], is that the frond is partly axial and partly foliar, almost in the sense in which this description

---

[1] Tieghem, P. van (1884).   [2] Theophrastus (Hort) (1916).   [3] Arber, A. (1919[1]).

can be applied to a phylloclade of one of the Rusceae. It is, in fact, a "short shoot" bearing a prophyll, while the buds (*b.* and *b.'*, fig. cxvii) produced to right and left in pockets (*p.* and *p.'*) at the base of the frond, are axillary to the prophyll, and develop, in their turn, into short shoots, each of which again gives rise to a prophyll, but to no other foliar member. The justification for this view is found in the comparison between the Lemnaceae and the reduced Aroid, *Pistia*; this comparison was suggested by Engler[1], and I have found, on making a detailed study of the *Pistia* shoot, that it is even more exact than he claims.

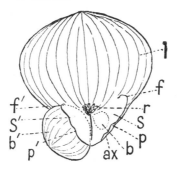

Fig. cxvii. *Spirodela polyrrhiza*, Schleid. A plant viewed from the underside (enlarged); *l.*, limb of leaf belonging to main axis, *ax.*; *r.*, roots (cut short); *p.* and *p.'*, pockets between wings of sheath (*s.* and *s.'*) and axis, enclosing buds, *b.* and *b.'*; *f.* and *f.'*, ligular flaps of sheath. [Arber, A. (1919[1]).]

Pistia Stratiotes, L., the River-lettuce of the Tropics, is a floating plant, which bears, on an abbreviated axis, a rosette of thick leaves, whose anatomy is of a type which suggests a phyllodic origin for the leaf-limb[2] (fig. cxviii, p. 150). This point does not, however, concern us now; the interesting question, from our present standpoint, is that of the relation of the *Pistia* shoot to the reduced shoot of the Lemnaceae. The series of sections through a vegetative growing point of *Pistia Stratiotes*, sketched in figs. cxix, 1—5, p. 151, shows that, in the River-lettuce, each leaf-limb (*l.*) is associated with a stipular sheath (*s.*), which, at its attachment to the axis, forms a lateral pocket (*S.p.*) on one side of the leaf-base; in this pocket a bud (*b.*) is produced. In the anomalous position of the buds, the leaf of *Pistia* thus corresponds to the frond of one of the Lemnaceae; it differs from it in having, on one side only, a pocket containing a bud, while in the Lemnaceae, there are bud-pockets on both sides. If, now, we imagine the shoot of *Pistia* so much reduced that it produces merely a single leaf (prophyll), which, however, is axillant to two buds instead of one, the resemblance is complete.

---

[1] Engler, A. (1877).   [2] See pp. 121, 125.

The entire assimilating system of the Lemnaceae, like that of *Ruscus*, I thus interpret as consisting of the prophylls borne by short shoots, but whereas, in *Ruscus*, these short shoots are borne on elongated axes of a lower order, in the Lemnaceae there are no long shoots, but the vegetative system is due to the development of successive generations of buds, which arise on the short shoots, and are axillary to the prophylls.

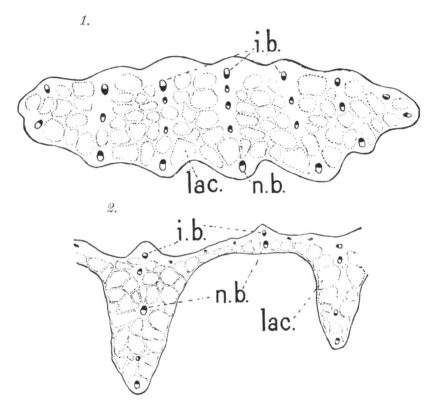

Fig. cxviii. *Pistia Stratiotes*, L. Transverse sections of leaf (× 21). Fig. 1, complete section near base of leaf. Fig. 2, two of the ribs in the median region of the fan-shaped limb; *n.b.* normally orientated vascular bundles; *i.b.*, inversely orientated bundles; *lac.*, lacuna. [Arber, A. (1919¹).]

(*d*) The "*Droppers*" of the Tulip and other Monocotyledons¹

A curious feature of the life-history of the Tulip is the lowering of the bulb into the soil, year by year, during the period of immaturity. This descent is accomplished by means of a tubular organ, the "dropper" or "sinker," which carries the terminal bud inside its tip. We may illustrate the first stages from the seedling of *Erythronium* which behaves similarly. Fig.

¹ Irmisch, T. (1850); Robertson (Arber), A. (1906); Döring, E. (1910); it should be noted that Döring makes the mistake of treating the new bulb as "axillary" to its own prophyll.

cxx, 7, p. 152, shows a plantlet whose cotyledon, *c.*, is still embedded in the seed, *s.*, while the plumular bud, *pl.*, instead of occurring in its usual position,

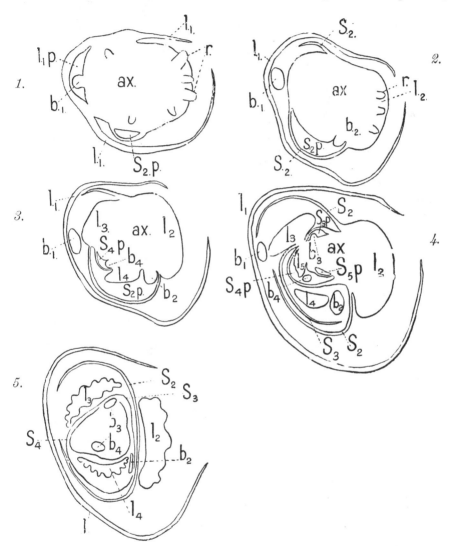

Fig. cxix. *Pistia Stratiotes*, L. Series of transverse sections through lateral bud, from base upwards (×14); *r.*, root; *ax.*, axis; l_1, first scale-leaf; l_2—l_5, limbs of successive foliage leaves; s_1—s_5, sheaths of these leaves; $l_1 p.$, $S_2 p.$, $S_3 p.$, $S_4 p.$, and $S_5 p.$, bud-containing pockets associated with these leaves, enclosing the buds marked b_1, etc. In fig. 5 all structures above leaf 4 are omitted. [Arber, A. (1919[1]).]

between radicle and cotyledon, is carried downwards by a dropper, *d.*, formed by a tubular prolongation of the cotyledon base. In the Tulip, all parts of the seedling finally perish, except the plumular bud, enclosed in an outer sheath

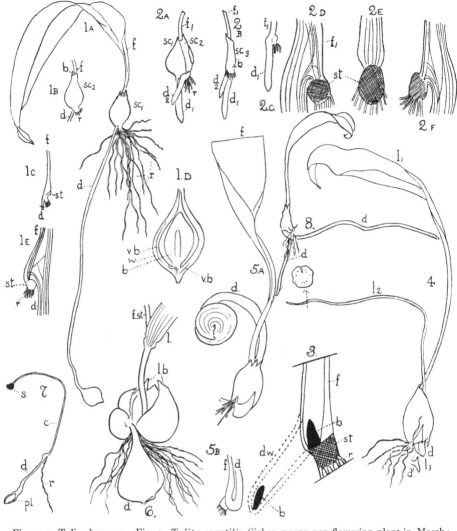

Fig. cxx. Tulip droppers. Fig. 1, *Tulipa saxatilis*, Sieber, young non-flowering plant in March; *f.*, foliage-leaf; sc_1, sc_2, scale-leaves; *b.*, axillary bud; *r.*, adventitious roots; *d.*, dropper; *st.*, stem. Fig. 1 A, whole plant (× ½); fig. 1 B, bulb with outer scale-leaf removed; fig. 1 C, bulb with inner scale-leaf and axillary bud removed, showing continuity of base of foliage-leaf and dropper; fig. 1 D, tip of dropper cut open (slightly enlarged), showing bulb, *b.*; wall of dropper, *w.*; vascular bundles, *v.b.*; fig. 1 E, junction of foliage-leaf, stem, and dropper, rendered transparent with carbolic acid, to show veining. Fig. 2, *Tulipa praecox*, Tenore, young, non-flowering plant in March; figs. 2 A—C, dissection of bulb; in fig. 2 B, the two outer scale-leaves, sc_1 and sc_2, are removed, and, in fig. 2 C, the third scale-leaf, sc_3, and the second dropper, d_2, which was connected with a rudimentary foliage-leaf. The main dropper, d_1, is a continuation of the foliage-leaf, f_1; *b.*, bud in axil of sc_2. Figs. 2 D—F, three views of junction of leaf, stem, and dropper in carbolic acid. Fig. 3, diagram to represent mode of origin of dropper of young *Tulipa*; the dotted lines show how the terminal bud, *b.*, instead of being left in its normal position on the stem (*st.*, crosshatched), is carried downwards by an outgrowth shaped like the finger of a glove, which develops from the leaf-base. Fig. 4, *Tulipa praecox*, Tenore, non-flowering plant in March; the foliage-leaf, l_1, terminates below in the dropper, *d.*; the incomplete foliage-leaf, l_2, has no dropper, but ensheathes a bud at its base; the foliage-leaf, l_3, which is a mere point, terminates below in the dropper, *d.'*. Fig. 5, *Tulipa praecox*, Tenore, non-flowering plant in March; fig. 5 A, plant as a whole; the dropper, *d.*, is abnormal in having passed upward and become coiled at the tip; fig. 5 B, shows the bend in the dropper when the bulb has been dissected so as to reveal the connexion between the leaf, *f.*, and the dropper, *d.* Fig. 6, *Tulipa* (garden var.), flowering plant in July, with a dropper, *d.*, from one of its lateral bulbs; *l.b.*, lateral bulbs without droppers; *f.st.*, flowering axis; *l.*, foliage-leaf. Fig. 7, *Erythronium grandiflorum*, Pursch, seedling; *d.*, dropper from cotyledon, *c.*, the tip of which is still embedded in the seed, *s.*; *pl.*, plumular bud; *r.*, primary root (× ½). Fig. 8, *Erythronium americanum*, Ker-Gawl., plant with two droppers, one of which is horizontal. [A.A., in part from Robertson (Arber), A. (1906).]

formed from the tip of the dropper. In the next season, the bud produces a single foliage-leaf, which is again prolonged at the base into a tubular dropper carrying the terminal bud[1] downwards into the soil. This phase of vegetative development, in which only one foliage-leaf is produced annually, is repeated for several years. Figs. cxx, I A—E, show an example of this stage for the case of *Tulipa saxatilis*, Sieber. The bulb is clothed with a scale, sc_1, which represents the remains of last year's dropper. On removing it, we come to another scale-leaf, sc_2 (fig. I B), and, when this is also removed, we reach the foliage-leaf, whose tubular base continues downwards into the dropper (figs. I C and E), which has the terminal bud in its tip (fig. I D). Such droppers may be very long; one of them, which I measured, belonging to *Tulipa sylvestris*, L., exceeded nine inches.

It may seem rather far-fetched to consider droppers under the heading of prophylls. My reason for doing so is that—although, in the young vege-tate plant, the dropper is formed from the annual foliage-leaf, to which the name prophyll cannot be applied—in the mature plants the droppers re-present a peculiar development of the prophylls of lateral shoots. And, even in the seedling where the dropper is cotyledonary, the term prophyll may perhaps be used, since the seed-leaf is related to the plumule, just as the prophyll is related to the lateral shoot. Fig. cxx, 6, shows a plant with an erect flowering axis, *f.st.*; a dropper corresponding to those of the immature non-flowering plants obviously cannot be formed in such a case, since the main axis, instead of remaining dwarfed, so that it can be carried down inside the base of the leaf which it produces, elongates and terminates in a flower. But here a dropper, *d.*, is formed by a lateral bud—that is to say, the first leaf (prophyll) of this bud carries its growing point a little way down into the soil. A similar occurrence of lateral droppers can be followed in figs. cxxi, I A—D, p. 154. This plant was non-flowering, and had one foliage-leaf, *a.*, which, when followed down, was found to be prolonged into a large normal dropper, *a.'*, enclosing the terminal bud (fig. I C). The prophyll of the bud, *e.*, is neither leafy nor dropper-like, while those of *b.* and *c.* are elongated upwards, but can scarcely be said to form droppers, since their downward extension, below the point of attachment, is very slight. The bud, *d.*, on the other hand, is enclosed in a prophyll (fig. I D), which only forms a small point when traced upwards, but which elongates downwards into a long dropper, *d.'*.

The Rusceae have given us examples of the fusion of the growing point of a shoot with its own prophyll, and the droppers of *Tulipa*, though super-

---

[1] In the paper cited, Robertson (Arber), A. (1906), I wrongly described the bud enclosed in the dropper of the young Tulip as *axillary*; it is, as Irmisch had long before pointed out, the *terminal* bud of the perennating main axis, which is carried downwards, step by step, by the successive leaves which it produces.

ficially so unlike the phylloclades of *Ruscus*, thus bear a morphological relationship to them.

In the genus *Erythronium*, the Dog's-tooth Violet, dropper-formation occurs on similar lines to that in *Tulipa*[1]. In *Gagea*, again, the lateral bud may be carried down by means of the prophyll, so that it is separated from

Fig. cxxi. Tulip droppers. Fig. 1, *Tulipa* (garden var.), June ; fig. 1 A, general view, with details shown in figs. 1 B—D (× ½); *a.*, foliage-leaf, connecting down to large normal dropper, *a.'* (fig. 1 C); *b.*, reduced foliage-leaf with bulb, *b.'*, enclosed in its sheathing base, but dropper region almost non-existent (fig. 1 B); *c.*, small foliage-leaf with dropper region, *c.'*, very short ; *d.*, foliage-leaf reduced to a mere point, but dropper, *d.'*, long (fig. 1 D); *e.*, foliage-leaf reduced completely, forming outer coat of bulb, but no dropper. In figs. 1 B—D, × indicates place of attachment to axis. Fig. 2, *Tulipa* (garden var.), June ; abnormal plant with shortly stalked bulb, *ax.*, in axil of lower foliage-leaf on aerial axis, *st.* (× ½). Fig. 3, *Tulipa* (garden var.); fig. 3 A, bulb with two droppers; figs. 3 B and C, dissections showing that the larger dropper, d_1, is the continuation of the base of the foliage-leaf, l_1, while the smaller dropper, d_2, terminates upwards in the rudimentary leaf, l_2 (× ½). [A.A.]

the parent axis, but in this genus the axillant leaf takes part in the stalk formation, and whereas in *Tulipa* the downward prolongation occurs on the ventral side of the leaf, in *Gagea* it is the dorsal side that is affected[2].

Dropper-formation is not confined to the Liliaceae. In the Ophrydeae (Orchidaceae) illustrated in fig. xiii, p. 27, the young tuber, consisting of a

---

[1] Blodgett, F. H. (1900).   [2] Irmisch, T. (1850), Pl. iv, figs. 20, 22, etc., and p. 142.

swollen root and an apical bud, is carried down into the soil by means of a "stalk[1]," which appears to be prophyllar. The occurrence of dropper forma-tion, in Monocotyledons so distinct from one another in affinities as the Orchids and Liliaceae, is an instance of the parallelism between the cohorts which we shall discuss in Chapter X.

The reversal of behaviour with regard to geotropism shown by these dropper-forming leaves, may be compared with the positive geotropism of the stolons of *Cordyline* (fig. xxvii p. 50). The resemblance of Tulip drop-pers to roots is not merely superficial, but may affect the actual mode of growth. I have observed, in a plant *Tulipa sylvestris*, L., by marking off the leaf and dropper into zones with Indian-ink, that, whereas the growth of the leaf above the attachment to the axis was, as usual in such cases, basal, the region of maximum growth of the dropper was in the zone immediately behind the apex. I think one would scarcely have anticipated that a prophyll would have gone to such lengths in the process of approximation to a root. It seems possible that some factor associated with underground life may so affect the chemistry of stems and leaves that they become liable to take on, not only some of the morphological features, but also some of the physio-logical characters of roots.

The examples of the prophylls of "short shoots" discussed in the pre-ceding sections of this Chapter illustrate a general trend observable elsewhere among Monocotyledons[2]—the tendency, namely, for the leaf to dominate the shoot, while the axis is relegated to a subordinate position.

[1] Irmisch, T. (1850), p. 142.   [2] See pp. 57—59.

CHAPTER VII

THE SEEDLING AND ITS SIGNIFICANCE

(i) INTRODUCTION

IN the later decades of the nineteenth century, the doctrine of recapitulation in animal embryology laid strong hold on the imagination of biologists, and inspired such classic work as the *Comparative Embryology* of F. M. Balfour. The dictum that "each organism in the course of its individual ontogeny repeats the history of its ancestral development," was not only easy to grasp and capable of picturesque presentation, but seemed to offer a royal, if arduous, road to the solution of many outstanding problems. Botanists as well as zoologists came under the influence of these ideas, and high hopes were entertained as to the phylogenetic revelations that might be expected if corresponding principles could be applied to plants. Attention was at first focused on the history of the embryo, from the moment of fertilisation up to the period of germination, and, from 1870 onwards, a mass of detailed work was produced on this subject, especially by Hanstein and his followers[1]. But, from the point of view of phylogeny, the results proved disappointing, and, towards the end of the last century, the centre of interest showed signs of shifting to those later phases of embryology represented by the first post-germination stages of the plantlet. The work then begun has indeed revolutionised our knowledge of seedlings, but it can scarcely be said, on the theoretical side, to have justified the anticipations of its pioneers, who looked to it, primarily, as a source of enlightenment on phyletic history. In the present chapter I propose to attempt some analysis of the hypotheses regarding the phylogeny of Monocotyledons, which have originated out of the study of seedlings. But before turning to the theoretical aspect of the subject, we must first pass briefly in review the different types of structure met with in seedlings of this class.

(ii) THE DESCRIPTION OF THE SEEDLING

A typical Monocotyledonous seedling consists of the primary root, and the hypocotyl, the latter bearing the single seed-leaf and the plumular bud. The primary root calls for no special remark: its existence is often ephemeral, its place being soon taken by adventitious roots. The hypocotyl, or axial region of the seedling, is frequently, though by no means always, much abbreviated; sometimes, indeed, it is so short as to be almost non-

[1] Sargant, E. (1914).

existent. In certain cases (e.g. *Arum maculatum*, L., fig. cxlv, 4 C, p. 190), it becomes swollen and tuberous after germination, while in others it is tuberous

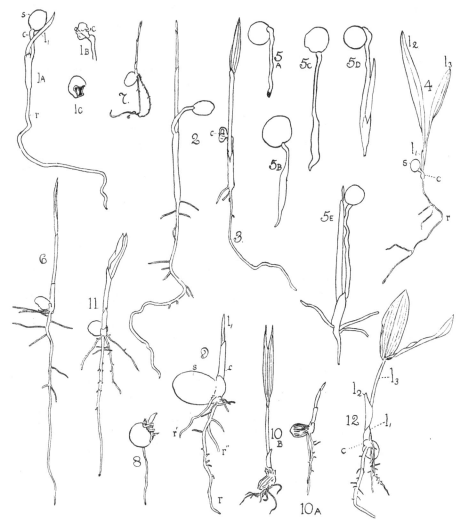

Fig. cxxii. Palm seedlings (drawings by Ethel Sargant): *c.*, cotyledon; *r.*, radicle; *r.'*, *r."*, adventitious roots; *l₁*, *l₂*, *l₃*, plumular leaves; *s.*, seed (all × ⅓). 1 and 2, *Chamaerops humilis*, L.; 1 A, whole seedling; 1 B, cotyledon dissected from shell to show bilobed form; 1 C, part of hard shell showing chambered structure. 2, older seedling. 3, *Trachycarpus* (*Chamaerops*) *Fortunei*, H. Wendl.; seed coats removed showing bilobed form of cot. 4, *Thrinax excelsa*, Lodd. 5, *Livistona* (*Corypha*) *australis*, Mart; 5 A—E, seedlings of successive ages. 6, *Pritchardia filifera*, Linden. 7, *Acanthophoenix crinita*, H. Wendl. 8, *Euterpe edulis*, Mart. 9, *Kentia Belmoreana*, C. Moore = *Howea Belmoreana*, Becc. 10 A and B, *Areca sapida*, Soland. = *Rhopalostylis sapida*, Wendl. et Drude. 11, *Desmoncus* "*minor*." 12, *Desmoncus* sp.

even in the ripe seed. In the group of aquatic families forming the cohort Helobieae, there is no endosperm, but reserves accumulate in the hypocotyl, which is hypertrophied to such a degree that the embryo—to which the term "macropodous" has been applied—is distorted out of all recognition[1]. Figs. cxxiii, 4—6, show transverse sections of the seed of *Zostera marina*, L. (Potamogetonaceae, Helobieae), in which the swollen hypocotyl (*hyp.*) enwraps the cotyledon (*c.*).

The cotyledon shows a remarkable range of variation which is best understood by comparison with the corresponding range of form met with

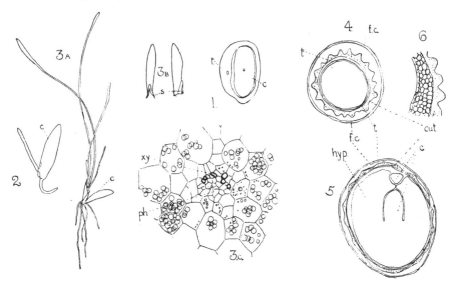

Fig. cxxiii. Figs. 1—3, *Aponogeton* sp.; fig. 1, transverse section of seed to show thick testa. *t.*, containing much starch, which is ultimately absorbed, and cotyledon, *c.*, with single vascular strand (× 7). Fig. 2, seedling (about nat. size); *c.*, green cotyledon; testa has rotted away. Fig. 3 A, older seedling, some of the leaves cut off short (× ⅓); *c.*, cotyledon; fig. 3 B, cotyledon removed and viewed from front and side to show sheath-wings, *s.* (slightly reduced); fig. 3 C, transverse section of the cotyledon bundle surrounded by starch-containing cells (× 77). Figs. 4—6, *Zostera marina*, L., Ryde, August. Fig. 4, transverse section of fruit; *t.*, testa; *f.c.*, fruit coat (× 23). Fig. 5, transverse section of fruit, higher up (× 23); *c.*, cotyledon, cut twice owing to its bent form; *hyp.*, enlarged hypocotyl. Fig. 6, transverse section of part of testa (× 47); the thick cuticular layer, *cut.*, represents the epidermis, but the individual cells are no longer distinguishable. [A.A.]

amongst Monocotyledonous foliage-leaves. In its completest and most typical development, the seed-leaf includes the following parts:

(i) basal sheath;

(ii) ligule, which may be closed and tubular, and is then conveniently distinguished as a ligular sheath;

(iii) limb, which, as in the case of the foliage leaf, I regard as equivalent to the petiole of a Dicotyledon[2]. The limb may be fairly uniform in structure throughout its entire length, or it may be differentiated into a proximal stalk and a distal region—more or less expanded —which corresponds to the pseudo-lamina of the foliage leaf.

---

[1] Arber, A. (1920[4]), pp. 248—9, 319.   [2] See also Bugnon, P. (1921[2]).

The diversity met with in the seed-leaves of Monocotyledons is largely due to the relative degree of development of the parts just enumerated: the basal sheath, or the ligule, may, for example, be entirely suppressed, while

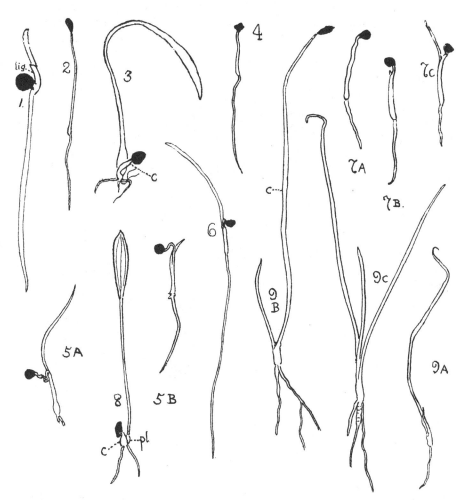

Fig. cxxiv. Seedlings of Liliaceae (nat. size); *pl.*, plumule; *l.*, plumular leaf; *c.*, cotyledon; *lig.*, cotyledon sheath. Fig. 1, *Littonia modesta*, Hook. [E.S.]. Fig. 2, *Allium Schoenoprasum*, L., var. *sibiricum* [E.S.]. Fig. 3, *Allium ursinum*, L. [E.S.]. Fig. 4, *Nothoscordum striatum*, Kunth [A.A.]. Figs. 5 A, B, *Ornithogalum fimbriatum*, Willd.[E.S.]. Fig. 6, *Hyacinthus amethystinus*, L. [E.S.]. Figs. 7 A—C, *Scilla hispanica*, Mill. [A.A.]. Fig. 8, *Lilium canadense*, L. [E.S.]. Figs. 9 A—C, *Galtonia Princeps*, Decne. [A.A.]

the limb may be extremely simple, or else complex and highly elaborated. A reference to a few of the examples here figured will show the degree of diversity produced. The seed-leaf of the Palm, *Chamaerops humilis*, L. (figs. cxxii, 1 and 2, p. 157), has a long basal sheath, a short ligule, and a limb,

which is stalk-like at the base, but has an expanded, bifid, distal apex, embedded in endosperm. The sucking part of the cotyledon may be developed
on a remarkable scale; that of the Coconut-palm (*Cocos nucifera*, L.) grows
so large, while gradually absorbing the endosperm, as to form a spongy,
deceptively fungus-like object, which finally comes to fill the entire cavity of
the nut[1]. The stalk-like part of the cotyledon may also be very conspicuous;

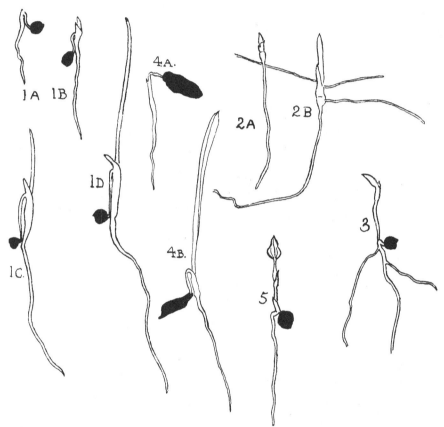

Fig. cxxv. Seedlings of Liliaceae (nat. size). Figs. 1 A—D, *Dasylirion* sp. Figs. 2 A, B, *Polygonatum biflorum*, Ell. Fig. 3, *Maianthemum Convallaria*, Wigg. Figs. 4 A, B, *Phormium tenax*, Forst. Fig. 5, *Smilax aspera*, L. [E.S.]

in *Hyphaene Petersiana*, Kl., it may exceed a foot in length[2]. One of the
Palms shown here—*Livistona* (*Corypha*) *australis*, Mart., figs. cxxii, 5 A—E,
p. 157—has a long-limbed seed-leaf, which does not, however, rival that of
Hyphaene. When the cotyledon in Palm seedlings is thus long-stalked, it
generally grows vertically downwards on germination, so that the radicle
and plumule are carried some depth into the soil (fig. cxxii, 5 E). In plants

[1] Wittmack, L. (1896).   [2] Karsten, H. (1847).

with very light seeds, on the other hand, the vertical elongation of the
cotyledon may have the opposite effect of raising the seed into the air.
This happens, for instance, in *Galtonia Princeps*, Decne. (fig. cxxiv, 9 B,
p. 159).

In many other Monocotyledons, the cotyledon stalk grows vertically
upwards from the seed, but the basal sheath bends downwards, carrying
with it the plumular bud. The result is that the stalk and sheath regions

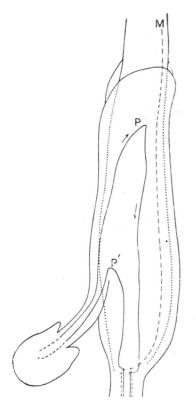

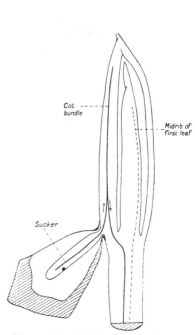

Fig. cxxvi. *Elettaria* seedling: diagram
showing limb, basal sheath and ligular
sheath of cotyledon; *P.*, *P.'*, cotyledon
bundles; *M.*, midrib of first plumular
leaf. [Sargant, E. and Arber, A. (1915).]

Fig. cxxvii. *Tigridia* seedling: diagram
showing limb (sucker) and ligular sheath
of cotyledon. [Sargant, E. and Arber, A.
(1915).]

of the cotyledon form a sharp angle with one another, and may even come
to lie almost parallel (e.g. *Dasylirion*, figs. cxxv, 1 A—D, and *Elettaria*, fig.
cxxvi). In *Elettaria* both ligular sheath and basal sheath are well developed,
but in *Tigridia* (fig. cxxvii) the basal sheath scarcely exists, and the cotyledon is reduced to a limb and a ligular sheath.

As an example of the contrast in the degree of development of the parts
of the cotyledon which may be met with within one family—Liliaceae—we

may take *Littonia modesta*, Hook. (fig. cxxiv, 1, p. 159), in which the limb of the cotyledon is very short, while the ligular sheath (*lig.*) is prominent, and *Galtonia Princeps*, Decne. (fig. cxxiv, 9 B), in which the ligule is wanting, but the cotyledon limb is extremely elongated.

Hymenocallis lacera, Salisb. (Amaryllidaceae, figs. cxxviii, 3 A—C) provides an example of a cotyledon with a basal sheath, *sh.*, which is considerably expanded and forms the outer coat of the first young bulb. The limb of the cotyledon is differentiated into a free stalk and a slightly enlarged distal region embedded in endosperm. The *Crinum* seedling shown in fig. iv, II A, p. 17, is of a similar type. Other seedlings of plants belonging to the Liliiflorae are shown in various chapters of this book; *Ruscus*, figs. cix,

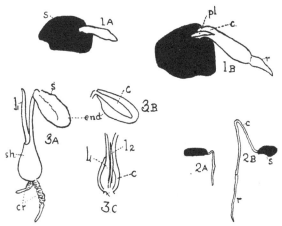

Fig. cxxviii. Seedlings of Amaryllidaceae (× ⅔); *s.*, seed; *end.*, endosperm; *c.*, cotyledon; *pl.*, plumule; l_1, l_2, plumular leaves; *r.*, radicle; *c.r.*, contractile roots. Figs. 1 A, B, *Crinum Macowani*, Baker [E.S.]. Figs. 2 A, B, *Chlidanthus fragrans*, Herb. [E.S.]. Figs. 3 A—C, *Hymenocallis lacera*, Salisb.; 3 A, external view of seedling, outer coat of seed removed; 3 B, endosperm split lengthways to show cotyledon; 3 C, young bulb cut lengthways to show first and second plumular leaves. [A.A.]

8 A—D, p. 140; cx, 20 A—C, p. 142; cxi, 25 A, p. 143; *Semele*, fig. cxiv, 47 A, p. 146; *Phormium*, figs. lxxxii, 21 A—C, p. 109; *Iris*, fig. lxxx, 2, p. 107; *Tamus*, frontispiece.

The seedling of *Aponogeton* (Aponogetonaceae, Helobieae, figs. cxxiii, 1—3, p. 158) differs from those hitherto considered, in the fact that the reserve for the young plant is stored in the cotyledon itself and not in the endosperm. The cotyledon is of the isobilateral equitant shape, so frequent among Monocotyledonous foliage-leaves (p. 106).

The seed-leaf of the Gramineae is sometimes treated as if it were fundamentally different in character from those of the other Monocotyledons, but, in reality, it conforms to the same general type, though it is peculiar in certain respects. The relation of the parts in the embryo and young seedling of the Oat (*Avena sativa*, L.) is shown in figs. cxxx and cxxxi, p. 164, and of the Maize (*Zea Mays*, L.) in fig. cxxxii A, p. 165 ; fig. cxxix shows the external appearance of various Grasses after germination, while the more detailed structure of the seedlings is indicated by some of the sections drawn in figs. cxxxiv, p. 167, and cxxxv, p. 169. I cannot here enter into all the conflicting views which have been held regarding the morphology of the Grass seedling, so I will content myself with briefly summarising the results of work published more fully elsewhere[1]. The scutellum (*sc.* in all figures), which is expanded and more or less elaborate in form, I regard as equivalent to the distal region of the sucking cotyledon in certain other Mono-

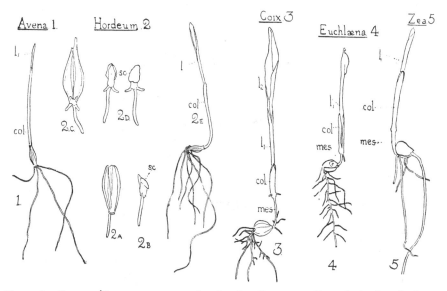

Fig. cxxix. Grass seedlings; *mes.*, mesocotyl; *col.*, coleoptile; *sc.*, scutellum; l_1, l_2, plumular leaves. Fig. 1, *Avena sativa*, L. ($\times \frac{1}{2}$). Figs. 2 A—E, *Hordeum vulgare*, L.; figs. 2 A and C, very young seedlings, roots only emerged; figs. 2 B and D, the same seedlings, but fruit coats and endosperm removed; A—D, slightly enlarged; fig. 2 E, older seedling ($\times \frac{1}{2}$). Fig. 3, *Coix Lacryma-Jobi*, L. ($\times \frac{1}{2}$). Fig. 4, *Euchlaena mexicana*, Schrad. ($\times \frac{1}{2}$). Fig. 5, *Zea Mays*, L. ($\times \frac{1}{2}$). [A.A.]

cotyledons; it may be compared, for instance, with the extremity of the seed-leaf in *Hymenocallis* (*c.* in fig. cxxviii, 3 B). A minor complexity may be introduced by the wrinkling of the secretory epithelium in contact with the endosperm, in a way that almost suggests the formation of a definite gland (fig. cxxxii C, p. 165)[2]. But the chief peculiarity of the seed-leaf of the Grasses is the behaviour of the proximal region of the stalk ; in some genera this region apparently suffers congenital fusion with the hypocotyl

[1] Sargant. E. and Arber, A. (1915) and Arber, A. (1923[3]).
[2] Sargant, E. and Robertson (Arber), A. (1905).

from the base of the ligular sheath downwards; its bundle, however, may
retain its independence (*sc.'* in fig. cxxxiii, p. 166 and *sc.b.'* in fig. cxxxv, 15 C,
p. 169. The result is that the scutellum is separated from the coleoptile (ligular

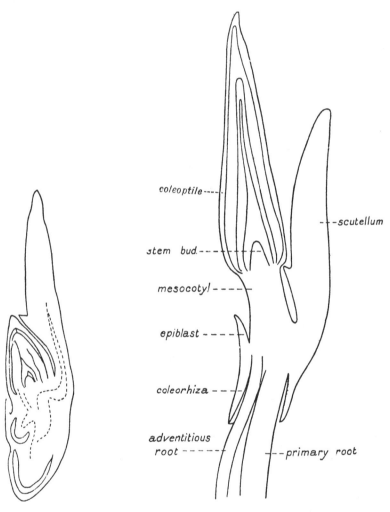

Fig. cxxx. *Avena sativa*, L.
Embryo in median section
(× 18). [Sargant, E. and
Arber, A. (1915).]

Fig. cxxxi. *Avena sativa*, L. Diagram of part of very
young seedling in median section. [Sargant, E. and
Arber, A. (1915).]

sheath) by a segment of axis + cotyledon-stalk—the so-called mesocotyl—
which may be of considerable length (*mes.* in figs. cxxix, 3, 4, 5, p. 163; cxxxi,
p. 164; cxxxii A, p. 165). A mesocotyl also occurs in certain Cyperaceae.
  Another peculiarity of the seed-leaf in the Grasses is that the basal

sheath is absent—a fact which may be correlated with the unusual degree of
development of the ligular sheath.

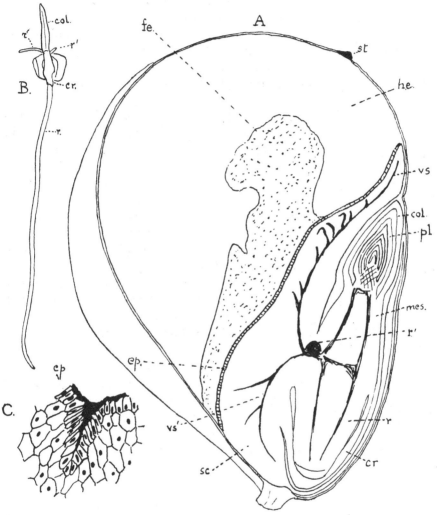

Fig. cxxxii. *Zea Mays*, L. A, diagram of the grain in radial longitudinal section, modified from the
figure in Sachs' Textbook; *h.e.*, horny part of endosperm; *f.e.*, floury part of endosperm; *st.*, stigma;
sc., scutellum; *ep.*, epithelium; *v.s.*, main vascular trunk of upper scutellum; *v.s.'*, one of the
vascular strands which ramify in the lower scutellum; *r.*, radicle; *r.'*, vascular tissue for first adventitious roots; *col.*, coleoptile; *cr.*, coleorhiza; *mes.*, mesocotyl. B, sketch of a seedling six days
old (nat. size). C, transverse section of part of margin of scutellum from a dry seed, showing
glandular infolding of secretory epithelium (× 137). [Sargant, E. and Robertson (Arber), A.
(1905).]

But it is not only the hypocotyl and seed-leaf of the Gramineae which
diverge from the usual Monocotyledonous type—the radicle also has its
own idiosyncrasy. It is enclosed in a sheath, the coleorhiza, which is con-

tinuous with the mesocotylar surface. In the light of the leaf-skin theory[1], it seems possible to interpret this sheath as a downward continuation of the

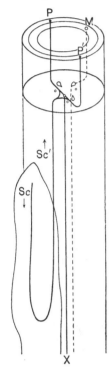

outer part of the cotyledon base. In other words, the cotyledon leaf-skin, which usually comes to an abrupt termination at the junction of hypocotyl and radicle, so that it may be visualised as a hollow cylinder open at the base, is, in the Grasses, prolonged to a blind end, so that it completely encloses the root-rudiment like the finger of a glove.

Another feature in the Grass seedling which has given rise to much speculation, is the occurrence, in certain genera, of a little non-vascular tongue-like organ—the epiblast (*bl.* in figures)—situated opposite to the scutellum. It is shown for the case of the Oat in figs. cxxx and cxxxi, p. 164; cxxxiv, 5 B, p. 167, and cxxxv, 14, p. 169 and for *Zizania aquatica*, L. in figs. cxxxv, 15 B and C. It has sometimes been interpreted as a second cotyledon, or it may be a mere outgrowth from the scutellum, corresponding, perhaps, to the basal auricles of the arrowhead leaf of *Sagittaria*, in a state of fusion. I think, however, that no solution yet propounded can be regarded as final, and that farther light is needed on the question.

Fig. cxxxiii. Diagram showing the vascular skeleton of scutellum, mesocotyl, and coleoptile (base only) in seedling of *Avena* type. [Sargant, E. and Arber, A. (1915).]

If we take the Grasses as demonstrating the degree of complexity to which a Monocotyledonous embryo can attain, it is to the Orchids that we must turn for examples of the opposite condition. Such minute, undifferentiated embryos as those of *Laelia* and *Bletia*, shown in figs. xiv E and F, p. 31, form a sharp contrast with the large and elaborate embryo of *Zea* (fig. cxxxii A, p. 165).

(iii) The Interpretation of the Seedling

Of all the structural differences between Monocotyledons and Dicotyledons, the most universal and the most striking is that divergence between the seedlings to which the two classes owe their names. But obvious as this difference is, there is no equally obvious answer to the question as to what is the morphological relation between the seedlings of the two groups —that of the Dicotyledon with its paired seed-leaves and that of the Monocotyledon with but one. This is a complex problem, and as a preliminary

[1] Saunders, E. R. (1922).

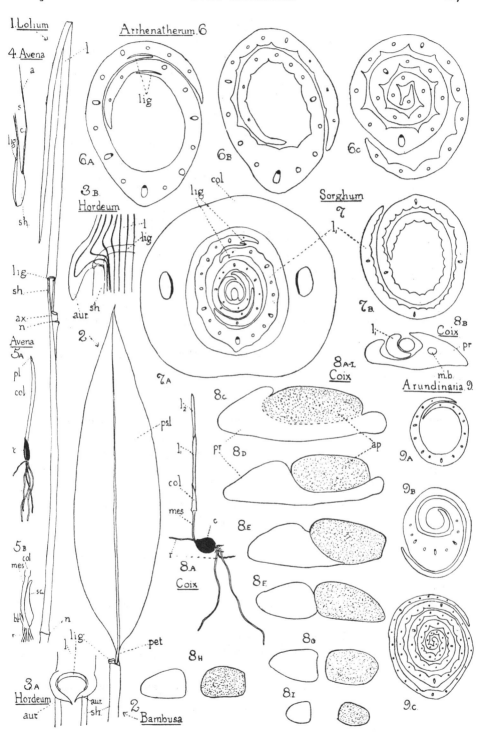

Fig. cxxxiv. Fig. 1, *Lolium multiflorum*, Lam.; foliage-leaf, showing sheath, *sh.*; ligule, *lig.*; limb, *l.*; axis, *ax.*; nodes, *n.* (× 0·5). Fig. 2, *Bambusa palmata*, Marl.; foliage-leaf, with sheath, *sh.*, of which part only is shown; ligule, *lig.*; petiole, *pet.*; pseudo-lamina, *ps.l.* (× 0·5). Fig. 3 A, *Hordeum Zeocriton*, L.; junction of limb, *l.*, and sheath, *sh.*, to show ligule, *lig.*, and auricles, *aur.* (about natural size); B, *Hordeum* sp.; diagrammatic sketch of one auricle, *aur.*, with part of junction of sheath and limb, to show vascular supply of auricle (enlarged). Fig. 4, *Avena strigosa*, Schreb.; lower palea to show awn, *a.* (consisting of column, *c.*, and "subule," *s.*), and ligule, *lig.* (slightly enlarged). Fig. 5, *Avena sativa*, L.; A, seedling; *pl.*, plumule; *col.*, coleoptile; *c.*, caryopsis; *r.*, roots (× 0·5); B, base of seedling shown in A, with endosperm and fruit coat dissected away to show scutellum, *sc.*, mesocotyl, *mes.*, and epiblast, *bl.* (slightly enlarged). Figs. 6 A—C, *Arrhenatherum avenaceum*, Beauv., var. *bulbosa*; transverse sections of leaf from apical bud; A, top of sheath, showing origin of ligule; B, origin of invaginations; c, limb (× 47). Figs. 7 A, B, *Sorghum vulgare*, Pers.; A, transverse section through coleoptile and plumular bud, showing origin of ligule in first leaf, l_1; B, l_1 at higher level (× 47). Figs. 8 A—I, *Coix Lacryma-Jobi*, L.; A, seedling about three weeks from sowing, from which sections shown in B—I were cut; *c.*, caryopsis; *mes.*, mesocotyl; l_1, l_2, plumular leaves; *r.*, roots (× 0·5); B, transverse section of bud in axil of first plumular leaf of seedling shown in A (× 77); the axillant leaf would lie toward the upper margin of page; *m.b.*, median bundle of prophyll, *pr.*; c—I, successive transverse sections of bud from axil of l_2 in A (× 193); prophyll is left white, and growing apex of bud is dotted; axillant leaf would lie toward upper margin of page. Figs. 9 A—C, *Arundinaria spathiflora*, Trin.; three transverse sections through a leaf, from sheath A to limb c (× 47). [Arber, A (1923³).]

to attacking it, it will be well to deal, in the first place, with the following relatively simple question, the answer to which will yield one of the elements necessary for the solution of the problem as a whole: *Is the seed-leaf of the Monocotyledon a single or a dual foliar member?* There are three sources of evidence to which we may look for help in finding an answer to this question—the external form of the seed-leaf; the anatomy of the seed-leaf; and the relation in form and structure between the seed-leaf and certain other juvenile leaves.

The first of the sources of evidence just enumerated—the external form of the cotyledon—cannot, in general, be said to give any indication of duality. The great majority, for example, of those Monocotyledonous seed-leaves represented in the figures illustrating the descriptive section of this chapter, can scarcely be claimed as showing any indication of doubleness in their external form. It is true that the basal region of the Monocotyledonous seed-leaf is very commonly completely closed and tubular, thus possibly suggesting a comparison with the tube formed in certain Dicotyledons by the union of the two cotyledon petioles by their margins (e.g. figs. cxxxvi, 6, 7, 11, p. 171). But this is merely a superficial similarity, from which no conclusions can be drawn, since a tubular basal sheath is extremely common among normal Monocotyledonous foliage-leaves, which no one can reasonably suspect of duality. When we turn from the sheath to the limb, we find that the only record among Monocotyledons of a seed-leaf whose limb is bifurcated, seems to be Sargant's[1] observation that in the Palms, *Chamaerops humilis*, L. and *Trachycarpus Fortunei*, H. Wendl., the sucking cotyledon embedded in the endosperm is deeply bifid (see figs. cxxii, 1 B and C and 3, p. 157). But it would be unsafe to deduce any theory of the seed-leaf from a peculiarity apparently confined to two related genera, when there are, on

[1] Sargant, E. (1903).

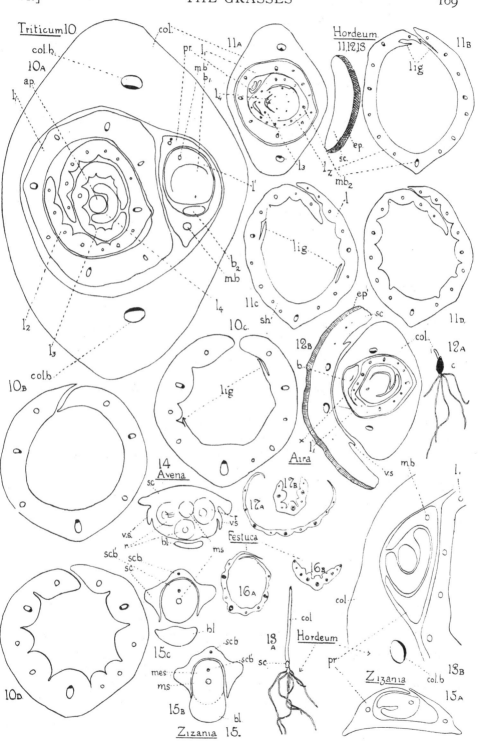

Fig. cxxxv. Figs. 10 A—D, *Triticum vulgare*,Vill. ; A, transverse section through seedling just above first node (× 47) to show, in axil of coleoptile, *col.*, a bud, b_1, whose prophyll, *pr.*, with median bundle, *m.b.*, has a bud, b_2, in its axil; *l.'*, first leaf of bud b_1, after prophyll; *m.b.'*, median bundle of *l.'*; l_1, l_2, l_3, l_4, plumular leaves; *ap.*, growing apex; B—D, sections from series through plumular leaf of another seedling (× 77); B, sheath ; C, origin of ligule ; D, limb. Figs. 11—13, *Hordeum vulgare*, L.; fig. 11 A, transverse section of seedling, just above insertion of coleoptile (× 23); *pr.*, prophyll of bud in axil of first leaf, l_1; *sc.*, scutellum; *ep.*, epithelium; figs. 11 B—D, further development of l_2; it will be seen from position of median bundle, *m.b_2*, that sections drawn in B—D are placed at right angles to that drawn in fig. 11 A; fig. 12 A, seedling (× 0·5) from which section drawn in fig. 12 B was cut; fig. 12 B, transverse section through seedling (just above insertion of coleoptile (but at lower level than fig. 11 A) to show bud, *b.*, in axil of coleoptile, connected by bridge marked with cross with first plumular leaf, l_1; scutellum shows columnar epithelium, *ep.*, ventral scale, *v.s.*, and two bundles (× 23); fig 13 A, seedling from which section shown in fig. 13 B was cut, drawn by Ethel Sargant, 13 days from sowing (× 0·5) ; endosperm removed to show scutellum, *sc.*; fig. 13 B, transverse section of bud in axil of coleoptile, *col.*, from seedling shown in fig. 13 A; *m.b.*, only bundle differentiated in prophyll, *pr.*; *col.b.*, coleoptile bundle (× 77). Fig. 14, *Avena sativa*, L.; transverse section through seedling, showing epiblast, *bl.*, free on one side, and scutellum, *sc.*, attached on other; *v.s.*, ventral scale of scutellum; *r.*, roots; *ms.*, mesocotylar stele (× 14). Fig. 15 A—C, *Zizania aquatica*, L.; A, transverse section of bud in axil of plumular leaf, which would lie toward upper margin of page (× 77); B and C, transverse sections of seedling close to base of epiblast, *bl.*, showing its attachment; scutellum, *sc.*; scutellum bundle, *sc.b.*; scutellum bundle in mesocotyl, *sc.b'*; mesocotylar stele, *m.s.* (× 14). Figs. 16 A, B, *Festuca* sp.; transverse hand sections of mature leaf (× 14); A, sheath; B, limb. Figs. 17 A, B, *Aira* sp.; transverse hand sections of mature leaf (× 23); A, sheath; B, limb. [Arber, A (1923³).]

the other hand, countless examples of similar sucking cotyledons which show no such bifurcation.

However, in the present connexion it is unnecessary to dwell at length on the external form of the Monocotyledonous seed-leaf, since those who have held that this organ is a dual structure, have based their view, not on its outward shape, but on its vascular anatomy. Our knowledge of the intimate structure of Monocotyledonous seedlings is mainly due to the work of Sargant, who made an extensive and detailed study of the structural relations of the seed-leaf, hypocotyl and primary root in the Liliaceae and other families—using microtome methods in the days before these had come to be generally employed in botany[1]. Her researches revealed the fact that a striking dual symmetry is a widespread anatomical feature of the single seed-leaf of Monocotyledons; there is, in other words, no median bundle, its place being taken by twin strands. Details must be sought in her papers, but a few examples may be cited here. In the case, for instance, of the Palm, *Chamaerops humilis*, L. (fig. cxxxvii, 1, p. 173) and the Amaryllid, *Agave spicata*, Cav. (fig. cxxxvii, 2), four bundles enter the cotyledon from the hypocotyl. In *Chamaerops* these four bundles are represented in the upper part of the cotyledon by eight, arranged in two groups of four (fig. cxxxvii, 1 B). In many other seedlings, two bundles are present in the cotyledon (e.g. Zingiberaceae; various Liliaceae, such as *Anemarrhena*, fig. cxxxvii, 3 E, and *Albuca*, fig. cxxxvii, 4); in others, again, there is a so-called "double bundle," with a single protoxylem group, flanked by two metaxylem and two phloem groups (e.g. *Cordyline indivisa*, Steud., fig. cxxxvii, 9; *Allium neapolitanum*, Cyr., figs. 5 B and C). In the plants in which a single bundle

[1] Sargant, E. (1902) and later papers.

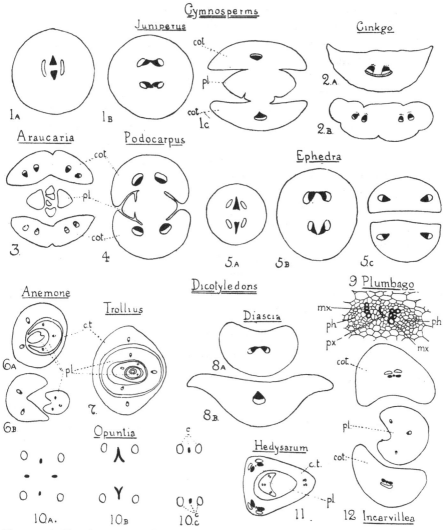

Fig. cxxxvi. Vascular structure of Gymnospermous and Dicotyledonous seedlings; *cot.*, cotyledon; *c.t.*, cotyledon tube; *pl.*, plumule. [Figs. 1—5, adapted from Hill, T. G. and Fraine, E. de (1908—1910).] Figs. 1 A—C, *Juniperus virginiana*, L.; transverse sections of seedling, from radicle, A, to base of cotyledon, C. Figs. 2 A, B, *Ginkgo biloba*, L.; transverse sections, lower, A, and higher, B, in one cotyledon; dots above xylem represent centripetal xylem. Fig. 3, *Araucaria Cunninghamii*, Forbes; transverse section just above cotyledon node. Fig. 4, *Podocarpus chinensis*, Sweet; transverse section at cotyledon node. Figs. 5 A—C, *Ephedra*; transverse sections from radicle, A, to cotyledons, C. Figs. 6 A, B, *Anemone apennina*, L.; transverse sections through cotyledon-tube and higher up (× 21) [Sterkx, R. (1900)]. Fig. 7, *Trollius europaeus*, L.; transverse sections through cotyledon tube (× 21) [Sterkx, R. (1900)]. Figs. 8 A and B, *Diascia Barberae*, Hook.; two sections of upper part of petiole of cotyledon, showing loss of doubleness in bundle in passing upwards [Lee, E. (1912)]. Fig. 9, *Plumbago micrantha*, Ledeb.; double bundle of cotyledon [Chauveaud, G. (1911)]. Figs. 10 A—C, *Opuntia Ficus-indica*, Mill.; vascular structure of seedling from radicle, A, to cotyledon strands, C. In 10 C, the double bundles, marked *c* and *c*, pass one into each of the cotyledons [Fraine, E. de (1910)]. Fig. 11, *Hedysarum coronarium*, L.; transverse section of cotyledon-tube and first plumular leaf [Compton, R. H. (1912)]. Fig. 12, *Incarvillea Delayvei*, Bur. et Franch.; transverse section through cotyledons and first plumular leaf (× 23). [A.A.]

supplies the cotyledon (e.g. *Zygadenus*, fig. cxxxvii, 6), there is often good reason to interpret this strand as equivalent to the pair of bundles met with elsewhere; for sometimes a single bundle in one species of a genus is replaced by two distinct bundles in a related form. In *Yucca aloifolia*, L., for instance, the cotyledon contains three bundles, the median one forming a midrib. The cotyledon of *Yucca gloriosa*, L., on the other hand, has a pair of bundles in the position occupied by the single median strand of *Y. aloifolia*. These, and many comparable observations which might be cited, indicate that a bundle system, in which there are twinned main strands instead of a single median bundle, is typical for the Monocotyledonous seed-leaf. A consideration of this peculiar vascular scheme, led Sargant to the view that the apparently single seed-leaf consists, in reality, of two foliar members, congenitally fused.

At the time that Sargant drew this conclusion, very little was known of the anatomy of Dicotyledonous seedlings. In the period—nearly a quarter of a century—that has elapsed since her theory of Monocotyledonous seedlings was propounded, a considerable bulk of work has been produced, dealing with the structure of the seed-leaf and hypocotyl in the Dicotyledons; much of this research owed its origin to the impetus which her writings gave to the study of the seedling. This later work has revealed the fact that a dual vascular symmetry is characteristic of the seed-leaves of Gymnosperms and Dicotyledons, as well as of Monocotyledons. The diagram of *Incarvillea Delavvei*, Bur. et Franch. (fig. cxxxvi, 12, p. 171), demonstrates the difference in vascular symmetry between the first plumular leaf (*pl.*), with its single midrib, and the cotyledons (*cot.*), in which the bundle shows a more or less patent doubleness. Fig. cxxxvi, 9, represents another cotyledonary "double bundle"—that of *Plumbago micrantha*, Ledeb., which may be compared with that of a Monocotyledon—*Cordyline indivisa*, Steud.— drawn in fig. cxxxvii, 9. In other Dicotyledonous or Gymnospermous seedlings illustrated in fig. cxxxvi, the cotyledon is traversed by two, or four, distinct bundles, so that the duality of the system is still more marked.

The actual meaning of the peculiar form of the cotyledon midrib, in which a more or less conspicuous doubleness prevails, will be considered a little later in this chapter; at the moment, we are only concerned with the fact that, since each cotyledon of the Dicotyledon proves to have a dual vascular symmetry—instead of the symmetry about a single midrib, which characterises normal foliage-leaves—the anatomical duality of the Monocotyledonous seed-leaf can no longer be claimed as an indication that the latter organ is equivalent to two foliar members.

It appears, then, that neither the consideration of external form, nor of internal structure, provides any convincing evidence of the bifoliar nature of the seed-leaf among Monocotyledons; it remains now to look at the

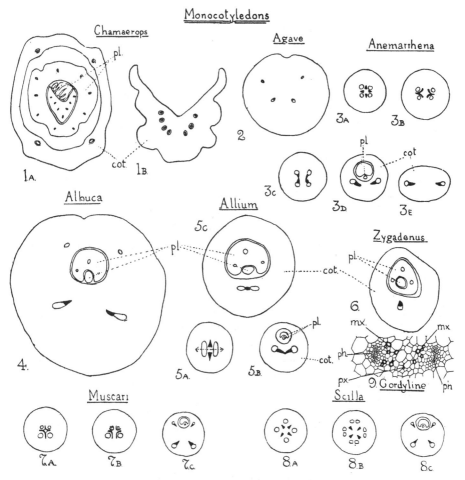

Fig. cxxxvii. Vascular structure of Monocotyledonous seedlings; *cot.*, cotyledon; *pl.*, plumule. Figs. 1 A and B, *Chamaerops humilis*, L.; transverse sections of seedling in Ethel Sargant's collection (× 14); fig. 1 A, sheathing region of cotyledon enclosing plumular bud; fig. 1 B, higher in cotyledon —bundles in two groups of four, but two of those in right-hand group enclosed in one sheath. Fig. 2, *Agave spicata*, Cav.; transverse section of cotyledon near base, from section in Ethel Sargant's collection (× 14). Figs. 3—8 [adapted from Sargant, E. (1902) and (1903)]. Figs. 3 A—E, *Anemarrhena asphodeloides*, Bunge; diagrammatic series of transverse sections through seedling from root, 3 A, to cotyledon, 3 E. Fig. 4, *Albuca Nelsoni*, N.E.Br.; transverse section (× 37) through base of cotyledon and plumular bud. Figs. 5 A—C, *Allium neapolitanum*, Cyr.; figs. 5 A and B, diagrammatic transverse sections through radicle, A, and base of cotyledon and plumular bud, B. The vascular tissue above arrow in 5 A, passes into the plumule, and that below the arrow, into the cotyledon; fig. 5 C, sketch of stage represented in 5 B, on same scale as fig. 4 (× 37). Fig. 6, *Zygadenus elegans*, Pursh; transverse section of base of cotyledon and plumular bud (× 37). Figs. 7 A—C, *Muscari armeniacum*, Leicht.; transition phenomena met with in one of the seedlings described by Sargant; A, root; B, hypocotyl; C, cotyledon enclosing plumular bud. The phloem groups continue straight up from root to cotyledon, while the xylem groups bifurcate and the halves fuse in pairs laterally. Fig. 8, *Scilla sibirica*, Andr.; A, root; B, hypocotyl; C, cotyledon enclosing plumular bud. The xylem groups are continued straight up from root to cotyledon—the protoxylems turning inwards during the passage—while the phloem groups bifurcate, and the halves fuse in pairs laterally. Fig. 9, *Cordyline indivisa*, Steud.; "double bundle" of cotyledon; *px.*, protoxylem; *mx.*, metaxylem; *ph.*, phloem. [Chauveaud, G. (1911).]

matter from a more general standpoint, and to study the cotyledons, not merely as individual organs, but in their relations as members of a shoot. Various botanists have been struck by the analogy between the *cotyledons* of the primary shoot, and the *prophylls* of secondary shoots; both cotyledons and prophylls are the first foliar members borne on an axis, and they agree in being often simpler in type than the mature foliage-leaves. The analogy is rendered more striking by the fact that, as we have shown (p. 131), in Dicotyledons there is usually a pair of laterally placed prophylls, while in Monocotyledons there is a single prophyll situated between the main axis and the lateral branch—that is to say, opposite to the axillant leaf. Now it has sometimes been maintained that the single prophyll of the Monocotyledon is, in reality, a dual structure, corresponding to the paired prophylls of the Dicotyledon; if this conclusion were accepted it would lend an additional touch of probability to the idea that the seed-leaf of the Monocotyledon is itself equivalent to the paired seed-leaves of the Dicotyledon. But a critical study of the structure and the general relations of the Monocotyledonous prophyll seems, on the contrary, to indicate that this organ is a single foliar member (see pp. 131—135), and hence the comparison with the prophyll fails to strengthen the case for the duality of the seed-leaf in this group.

And now, having considered the external form and internal structural relations of Monocotyledonous seed-leaves, and having compared them with prophylls, we are, I think, justified in drawing the conclusion that *the seed-leaf of Monocotyledons is a single foliar member*. If this result may be regarded as established, we can pass on to the broader problem of the relation of the seed-leaves of Dicotyledons and Monocotyledons. The conclusion that the seed-leaf of the Monocotyledon is a single member, must necessarily rule out Sargant's theory that it is equivalent to the two cotyledons of the Dicotyledon in a state of congenital fusion. This theory was, however, so widely based, and so admirably worked out, that it deserves farther study: if we can carry analysis to the point of detecting exactly where and how it breaks down, we ought to gain a clearer insight into the significance of seedling structure. In treating of the anatomy of the cotyledon, I have drawn attention to some of Sargant's observations, but her theory was based upon the behaviour of the cotyledon strands in the transition region between seed-leaf and root, rather than upon the structure within the cotyledon itself. An example on which she laid great stress was the type of vascular symmetry found in the seedlings of the Liliaceous genera *Anemarrhena, Galtonia* and *Albuca*; we will briefly recall her account of the transition phenomena in these seedlings—following the strands, according to her convention, from above downwards. In these genera the cotyledon is supplied by two massive bundles, which, above the plumular bud, occupy the foci of the elliptical transverse section (fig. cxxxvii, 3 E, p. 173). Below the plumular bud they meet to form the stele of the hypocotyl; as they approach one another, each trace opens out into a double bundle (fig. 3 C). Before the double traces meet, the xylem of each branches in three directions, and for a very short distance there are six protoxylem rays and four phloem groups in the stele (fig. 3 B). Lower down, the stele becomes root-like and tetrarch by the union of two pairs of adjacent lateral protoxylems (fig. 3 A). Sargant pointed out that an almost identical type of skeletal system was to be found in *Eranthis*, one of the Ranunculaceae, and more recent work has revealed its occurrence in species belonging to the Berberidaceae, Lauraceae, Cactaceae (e.g. *Opuntia*, fig. cxxxvi, 10 A—C, p. 171), and Bignoniaceae. In these Dicotyledons the vascular strand contributed by each of the two seed-leaves plays precisely the same part in the formation of the root-stele as one of the two strands of the cotyledon of *Anemarrhena*. Described in these terms, the similar behaviour of the hypocotyl strands in these Liliaceae and Dicotyledons, may appear, at first sight, to lend colour to the view that each of the bundles in the cotyledon of *Anemarrhena* represents a distinct seed-leaf. Such an interpretation of the facts seems to me, however, to be inadmissible. The significance of the vascular duality in the cotyledon, and also of the behaviour of the strands in the hypocotyl, has, I think, been misunderstood, simply because Sargant, and certain later workers who have studied the anatomical structure of the transition region in seedlings, have traced the bundles from above downwards—as in the description cited above—and have hence been led into the fundamental error of treating the root-stele as *made up of* cotyledonary and plumular strands. De Candolle, in 1827, laid down the principle that each organ should be considered "comme se développant ou sortant de celui qui lui sert de support immédiat," and he pointed out that the business of the morphologist was "d'étudier les exsertions et non les insertions." It seems to me that any attempt to reverse this procedure is as foredoomed to failure as an attempt to build a house from the chimney-pots downwards. We cannot logically treat the vascular system of the root as composed of foliar strands. We should rather look to root structure for an explanation of the ground-plan and anatomy of the cotyledonary members: and as soon as we do so, our difficulties vanish.

Primary roots throughout the vegetable kingdom are commonly diarch or tetrarch, and I wish to suggest that the opposite phyllotaxis of the first two leaves—whether they occur at the same node (cotyledons of Dicotyledons), or successively at two nodes (cotyledon and first plumular leaf of Monocotyledons)—is due to the bilateral symmetry impressed upon the shoot by the root with which it is in continuity. In anatomy, also, these early leaves hark back, as it were, to root structure. The root-bundle shows typically what Chauveaud[1] calls the "disposition alterne"—a proto-

---

[1] Chauveaud, G. (1911).

xylem strand separating two phloem strands. The "double bundle," which so commonly occurs in place of the midrib in the seed-leaves of Dicotyledons, is essentially the upward continuation of such a root-bundle. This view, though differing somewhat from that of Chauveaud, seems to me to be a natural deduction from his observations. Lenoir[1], who has studied seedlings on the lines laid down by Chauveaud, but whose work is more critical from our present standpoint, since it is based on microtome series rather than on hand-sections, explicitly takes the view that the cotyledon strand is the continuation of the root bundle, with which, however, its identity is not absolute, since it is modified to a greater or less extent by the addition of new tissues to its upper region. There is little doubt that this generalisation may be accepted for Dicotyledons; Monocotyledons do not, however, conform to it universally. The variety in the skeletal system of their seedlings, revealed by Sargant's work, provides an astonishing contrast to the stereotyped scheme found among Dicotyledons. This is probably due in part to the internodal abbreviation, characteristic of Monocotyledons, which brings the strands of the primary root into more direct association with the plumular traces than is usual among Dicotyledons. In part it may also be due to the breadth of the sheathing cotyledonary base, which allows space for the passage of minor bundles from the root to the cotyledon, in addition to the main strands. But, whatever the causes may be, the exceptions met with among Monocotyledons prevent us from treating Lenoir's law as of universal application: it must be admitted that, though in some cases it explains the peculiarities of the Monocotyledonous seed-leaf, in others it is invalid. I should like to suggest the following interpretation for those cases in which it fails.

Nearly all the seedling roots in this class are diarch or tetrarch—that is to say, they present *a dual symmetry*. In some instances the seed-leaf is supplied by strands representing the *whole* vascular system of the root, and it thus inevitably receives the impress of the dual symmetry of that organ. It follows that the anatomical duality of the Monocotyledonous seed-leaf does not always mean the same thing, since these seed-leaves fall into two categories, both of which are liable to many modifications in detail. In one type the doubleness is due to the fact that the cotyledon receives either a pair of bundles, or else one "double bundle," representing the upward continuation of one root pole (e.g. *Allium neapolitanum*, Cyr., figs. cxxxvii, 5 A—C, p. 173); its duality is thus precisely equivalent to that of each seed-leaf in a Dicotyledon. But in the other type the cotyledon receives two bundles, each of which represents half the entire root-stele (e.g. *Anemarrhena*, figs. cxxxvii, 3 A—E); each of these two bundles is therefore equivalent

---

[1] Lenoir, M. (1920).

to the double bundle of a Dicotyledon. But this explanation, if accepted, does not in any way affect our main contention—namely that, both in Mono-cotyledons and in Dicotyledons, the anatomical divergence between cotyledons and later foliage leaves is due to the fact that the former are necessarily supplied by root strands and the latter by stem strands. Coty-ledons have a strong tendency to dichotomy, which has been interpreted as an archaic ancestral character[1], but I am inclined to think that it is merely a mechanical result of their "root bundle" equipment.

It is interesting to find that "cotyledonary" anatomy is not absolutely confined to the cotyledons, but may be shared by a succeeding leaf, if it happens to draw its vascular supply direct from the root. Compton[2], in his fine work on the anatomy of Leguminous seedlings, has shown, for instance, that in the case of *Pisum sativum*, L., which has a triarch root, two of the xylem poles, each accompanied by a group of phloem to right and left, pass into the cotyledon, thus providing a double bundle for each, while the third root pole passes into the first plumular leaf, which is a reduced trifid struc-ture. It is true that, in this third bundle, the appearance of doubleness is lost by fusion before the strand enters the leaf, but this does not alter the fact that the anatomical structure of this primordial leaf is essentially that of a cotyledon. Probably *Allium neapolitanum*, Cyr., described by Sargant, is a comparable case, though here the first plumular leaf is relatively belated in appearance, and the root character of its median strand is obliterated be-fore the bundle leaves the axis.

The fact that the structure of cotyledons and of certain juvenile leaves is, in reality, "root anatomy," puts a somewhat different complexion upon the comparison with the prophyll which we considered earlier in this Chapter. No doubt prophylls are, in a broad sense, homologous with cotyledons; both are the first leaves of vegetative shoots—the shoots only differing in the fact of being lateral in one case and primary in the other. But the skeletal system is wholly different in these two foliar types, since prophylls are supplied by stem strands, and cotyledons by root strands. The binerved structure of the Monocotyledonous prophyll, though at first glance recalling the dual symmetry of the Monocotyledonous seed-leaf, is, as already shown, a fundamentally different thing (pp. 131—135); in the prophyll, it usually de-pends on an exaggerated development of one or both of the main laterals, while in the seed-leaf, it is frequently due to a farther separation of the components of the double bundle derived from the root. For it often happens that the protoxylem is seen to die out completely as the root pole is followed upwards, leaving two groups of metaxylem associated with two phloem groups. In

---

[1] Bugnon, P. (1922); see also Bugnon, P. (1923) for the development of a theory of seedling anatomy essentially different from that here maintained.

[2] Compton, R. H. (1912).

other words, the root bundle becomes, in its upward course, transformed into a pair of bundles. In certain Monocotyledons, the bundle derived from one pole of a diarch root passes up into the seed-leaf, while the other is responsible for the vascular supply of the plumule (e.g. *Allium neapolitanum*, figs. cxxxvii, 5 A—C, p. 173). But in such a seedling as that of *Anemarrhena* (fig. cxxxvii, 3), we have two double bundles passing up into the single cotyledon, whereas, in the *Anemarrhena*-like Dicotyledons (e.g. *Opuntia*, fig. cxxxvi, 10, p. 171) one of these double bundles supplies each cotyledon. Sargant treated this as an indication that the cotyledon of *Anemarrhena* represents the two seed-leaves of these Dicotyledons in a state of fusion, but it seems to me that, if we trace the strands upwards instead of downwards, a much simpler explanation at once presents itself. In *Anemarrhena* the cotyledon is a fully developed organ at a stage at which the plumule is too embryonic to receive strands from the primary root, so all the bundles composing the root-stele pass upwards into the sheathing cotyledon—simply because there is nowhere else for them to go. In the case of *Anemarrhena*, the root is tetrarch, but the cotyledon, instead of being supplied, in the simplest possible manner, by four bundles derived from the four root poles (as in the seedlings shown in figs. cxxxvii, 7 and 8) receives two bundles only, derived from those of the opposite poles, reinforced by contributions from the alternating poles. This complication does not, however, affect our argument. In the case of those Dicotyledons in which the transition phenomena in the hypocotyl recall those in *Anemarrhena*, the same vascular apparatus which in the latter genus passes into the single cotyledon with its completely ensheathing base, is shared between the two paired cotyledons (fig. cxxxvi, 10, p. 171). But this is a mere mechanical necessity, and in no way proves that the two seed-leaves of the Dicotyledon are together equivalent to the single cotyledon of the Monocotyledon.

If the seed-leaf of the Monocotyledon is admitted to be a single foliar member, it must be homologous with one of the seed-leaves of a Dicotyledon. Assuming a dicotylar ancestry for both classes, A. W. Hill[1] has put forward the interesting theory that the monocotylar condition may have arisen by division of labour between the two seed-leaves, one having retained the cotyledonary position and function, while the other became modified to form the first plumular leaf. He shows that, in certain geophilous species of the Dicotyledonous genus *Peperomia*, such a differentiation between the cotyledons can actually be observed; one of them leaves the seed and becomes epigeal and assimilating, while the other remains permanently in the seed as a sucker. It is, of course, conceivable that the difference between the seedlings, which distinguishes Monocotyledons, as a class, from Dicotyledons, may have come about in this way, but I am inclined to think that

[1] Hill, A. W. (1906).

the hypothesis is a superfluous one. If we look at the matter broadly, *the question arises whether the postulate of the absolute equivalence of the cotyledonary apparatus in Monocotyledons and Dicotyledons is really so inevitable as is generally assumed.* I have myself come gradually to the conclusion that no logical necessity underlies this postulate, and that it has been taken for granted because botanists have been hypnotised by their own terminology. Having invented the name "cotyledons" for these members, they have come to look upon them almost as organs *sui generis*, and to forget that they are merely the earliest leaves of the plant. Students—and teachers also—slip easily into the habit of visualising the first *plumular* leaf as the first leaf of the seedling, despite the fact that, before the end of the eighteenth century, Goethe had made it clear that the cotyledons were the first leaves borne by the axis, and that their level of attachment was its first node[1]. The use of the label "prophyll" for the first leaf of a lateral shoot, encouraged a parallel misconception. As Duval-Jouve[2] pointed out, nearly half a century ago, the words "Vorblatt" and "préfeuille" (with which our term "prophyll" may be bracketed) are peculiarly unfortunate in that their etymology suggests an organ *preceding* the leaves, rather than the first leaf borne on an axis. He felt this drawback so keenly that he coined the substitute "primefeuille," which emphasises the character of the prophyll, as itself the first leaf of the shoot. But botanists have never been quick to recognise the importance of delicate adjustment of language to ideas, and Duval-Jouve's suggestion seems to have passed unheeded.

I think it is this tendency to treat cotyledons, and their homologues, prophylls, as organs apart, which is responsible for the widespread expectation that the second seed-leaf of the Dicotyledons is waiting to be discovered among Monocotyledons—either as a rudiment, or fused with its fellow, or masquerading as the first plumular leaf. *But why should it be there at all?* Herb Paris has its foliage in a four-membered whorl, while the related Trinity-flower has a whorl of three, but no one argues that *Paris* is derived from *Trillium* by splitting of one of its three leaves, or that *Trillium* comes from *Paris* by fusion of two of the leaves, or reduction of one to a rudiment, or by displacement of one to another node. Why should we not, then, look at the cotyledonary system from the same standpoint, and suppose that one section of the Flowering Plants is monocotylar—not because there has been a fusion of two seed-leaves, nor because there has been suppression or displacement of the second—but because the growth-rhythm happens to be of the type which produces a single leaf at the first node? Nothing could be more natural than such an occurrence in a group of plants in which the leaf-bases have a marked tendency completely to ensheathe the axis—a character which, in itself, absolutely precludes the production of two leaves at a node[3].

[1] Goethe, J. W. von (1790).   [2] Duval-Jouve, J. (1877).   [3] Bugnon, P. (1921[3]), p. 502.

CHAPTER VIII

THE REPRODUCTIVE PHASE

THE study of the flower and fruit of Monocotyledons is a world in itself, and it would be impossible to do justice to it in the few pages at my disposal. Fortunately there are numerous books devoted to the subject to which the reader may be referred. Of the more exhaustive treatises it is only necessary to name Engler and Prantl's *Die natürlichen Pflanzenfamilien*, Engler's *Das Pflanzenreich*, and Baillon's *Histoire des Plantes*, while, as a handbook, there is A. B. Rendle's small convenient volume, *The Classification of Flowering Plants*, Vol. I, which may be supplemented by the use of J. C. Willis's *Dictionary of the Flowering Plants and Ferns*. The existence of these and other books, in which the flower and associated parts receive full consideration, absolves me from making any attempt at a comprehensive account of the reproductive shoot of Monocotyledons, so I propose to confine myself to a few special topics.

A feature in which the inflorescence of Monocotyledons often differs from that of the other Flowering Plants, is the presence of a *spathe*. The spathe is a modified leaf borne on the inflorescence axis; it generally corresponds to the prophyll of the vegetative shoot, and is then distinguished by the name *bracteole*. It is often a relatively large structure, so that, at least in the early stages, it completely encloses the flowers. The spathe of *Iris*, illustrated in figs. cxxxviii, 4 A—C, is a typical example of an inflorescence-prophyll. Fig. 4 A shows a leaf, *ax.l.*, borne on the inflorescence axis, i_1, and axillant to the lateral inflorescence axis, i_2. The first leaf produced by i_2 is the prophyll, *sp.*, which is traversed by two principal veins with a transparent membrane between them; the attenuated non-vascular texture of the median region of the spathe is doubtless—as in the case of similar features in the prophylls of vegetative shoots (p. 131)—to be attributed to development under pressure.

I am inclined to think that the great frequency of occurrence and the large development of Monocotyledonous spathes is correlated with the conspicuous part played by the basal sheath in Monocotyledonous foliage leaves; for, in various genera evidence may be found that the spathe is a leaf-base structure. In our British Ramsons, *Allium ursinum*, L., there is, in normal plants, a long peduncle, bearing a spathe formed of two connate bracteoles, *sp.*, which enclose the inflorescence (fig. cxxxviii, 1 A), but in the abnormal example shown in fig. 1 B, while one of the spathe leaves (*sp.*) is of the usual

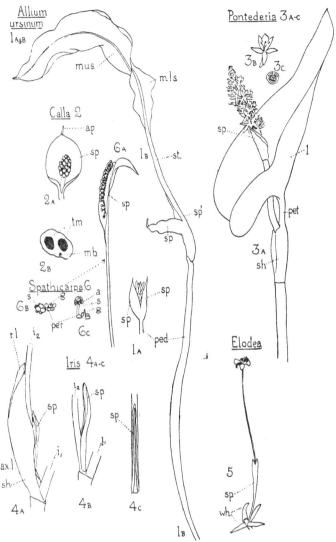

Fig. cxxxviii. Spathes. Figs. I A and B, *Allium ursinum*, L.; fig. I A, normal inflorescence with spathe consisting of two connate leaves, *sp.* and *sp.*, borne by peduncle, *ped.*, of which only apical region is shown (× ⅓); fig. I B, abnormal inflorescence, consisting of unusually long and thick peduncle, bearing two connate spathe leaves, one of which, *sp.*, is normal, while the other, *sp.'*, has grown out into a foliage leaf with a sheathing base, identical with the normal foliage leaves of the plant. The blade is inverted by a twist in the region of the petiole marked *st.*; *m.l.s.*, morphologically lower surface; *m.u.s.*, morphologically upper surface. The flower buds were enclosed in the spathe as usual, but are not visible in the drawing (× ⅓). Lyme Regis, April, 1923. Figs. 2 A and B, *Calla palustris*, L.; fig. 2 A, spathe, *sp.*, solid apex, *ap.* (× ⅓); fig. 2 B, transverse section of spathe apex; *m.b.*, midrib; *t.m.*, lateral tracheal masses (× 9, *circa*). Figs. 3 A—C, *Pontederia rotundifolia*, L.; fig. 3 A, reproductive shoot bearing leaf with sheath, *sh.*, petiole, *pet.*, and limb, *l.*, followed by spathe leaf, *sp.*, reduced to sheath region (× ⅓); fig. 3 B, single flower; fig. 3 C, transverse section of ovary; front cavity contains seed, while the other two are empty and sterile. Figs. 4 A—C, *Iris* sp.; fig. 4 A, base of an inflorescence axis, i_1, bearing leaf, *ax.l.*, consisting of sheath, *sh.*, and reduced limb, *r.l.*, in whose axil arises the lateral inflorescence axis, i_2, bearing the prophyllar spathe, *sp.* (× ⅓); fig. 4 B, structures shown in fig. 4 A, with leaf, *ax.l.*, removed; fig. 4 C, spathe from adaxial side. Fig. 5, *Elodea canadensis*, Mich., female flower with spathe, *sp.*, arising in the axil of a leaf of the fertile four-leaved whorl, *wh.* (about natural size). Wells, Norfolk, 1921. Figs. 6 A—C, *Spathicarpa sagittifolia*, Schott; fig. 6 A, inflorescence, peduncle incomplete, to show spadix fused with spathe, *sp.* (reduced); figs. 6 B and C, views from top and side of one synandrium, *s.*, with anthers, *a.* (shaded), and one female flower, with gynaeceum, *g.*, and perianth members, *per.* (dotted). [A.A.]

form and size, the other (*sp.'*) has grown out into a typical foliage leaf, with a well-developed stalk, terminating in the characteristic inverted limb. The comparison of *sp.* and *sp.'* undoubtedly confirms the idea that the spathe leaves are leaf-base structures. The spathe of *Pontederia rotundifolia*, L. (*sp.*, fig. cxxxviii, 3 A), is also clearly equivalent to the leaf-base (*sh.*) of the fully developed foliage leaf which precedes it. In Palms and Aroids (e.g. *Arum maculatum*, L., fig. cxlv, p. 190, and *Sauromatum guttatum*, Schott, fig. clviii, p. 208), the spathe is a particularly noticeable feature, and sometimes attains gigantic proportions. Long ago the spathe of the Araceae attracted the attention of Sir Thomas Browne, who dwelt upon "the purple pestil of Aaron, and elegant clusters of dragons, so peculiarly secured by nature, with an umbrella, or skreening leaf about them." The spathe of *Arum*

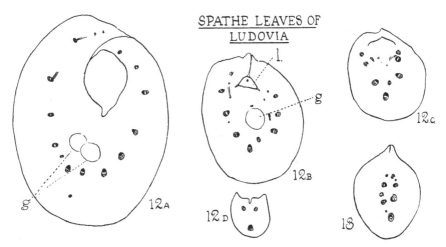

Fig. cxxxix. *Ludovia crenifolia*, Dr.; transverse sections through apical region of spathe leaves (× 14). Figs. 12 A—D, series from one spathe. Fig. 13, section from tip of another spathe; *g.*, gum-canals; *l.*, ? ligule. [Arber, A. (1922[1]).]

maculatum is so broad at the base that its attachment surrounds the axis more than once, forming a "wrap-over" (fig. cxlv, 2 C, p. 190). In *Calla palustris*, L. (fig. cxxxviii, 2 A), the spathe is small in size, but it is interesting as showing the solid tip (2 B) which is not infrequently characteristic of these sheathing structures, and which may, I think, be explained as a vestigial petiole. In *Ludovia crenifolia*, Dr. (Cyclanthaceae), the anatomy of the radial apical region is in harmony with the petiolar interpretation, for it suggests that of an *Acacia* phyllode (fig. cxxxix, 13). The importance of the spathe perhaps reaches its culmination in such an Aroid as *Spathicarpa* (fig. cxxxviii, 6 A) in which the flower-bearing part of the axis (spadix) is completely fused with its prophyll, so that the flowers appear to be growing out of the midrib of the spathe.

When we turn to the study of the individual flowers of Monocotyledons, we find, as in Dicotyledons, a complete range from the more generalised hermaphrodite types with numerous parts, to reduced and sometimes unisexual forms. The Alismaceae in certain respects show generalised characters,

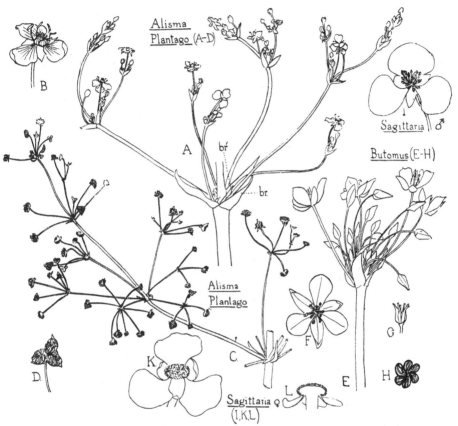

Fig. cxl. Flowers of Alismaceae and Butomaceae. A—D, *Alisma Plantago*, L., Chiddingly, Sussex, Sept.; A, one node of inflorescence (× ½); *br.*, bract; *br.'*, bracteole; B, single flower of A, enlarged; C, part of one node of infructescence (× ½); D, one fruit enlarged to show numerous achenes and remains of sepals; E—H, *Butomus umbellatus*, L., Fordham, Cambs., Aug.; E, inflorescence (× ½); F, flower from above to show 6 carpels and 9 stamens (slightly reduced); G, carpels from faded flower (× ¼); H, transverse section across basal region of carpels in G, to show ovules scattered on carpellary walls (× about 1); I—L, *Sagittaria montevidensis*, Cham. et Schlect. (× ½); I, male flower; K, female flower; L, female flower, with outer perianth members removed, cut longitudinally. [A.A.]

for they include plants in which both stamens and carpels (*Sagittaria*, figs. cxl, I, K, L) or the carpels alone (*Alisma Plantago*, fig. cxl, D) are numerous, hypogynous and free. The floral type in some degree recalls that of the Ranales. The mode of life of the Alismaceae is commonly aquatic—a characteristic which they share with many other Helobieae; but since I have dealt with them elsewhere[1] from this point of view, I will only now recall that the Water-plantain, *Alisma Plantago*, L., fails to flower if it finds itself in deep water, and then produces only ribbon-leaves (fig. cxli). The flowers of some members of the family are more specialised than those of the *Alisma* type; the Arrowhead, *Sagittaria*, for instance, has separate male and female flowers (fig. cxl, I, K, L). The same character is repeated in another aquatic Monocotyledon, *Sparganium*, the Bur-reed (fig. cxlii), in which, however, each male flower has only two to three stamens, and each female flower, a single carpel. Indeed those cases, in which the androecium and gynaeceum have numerous members—though possibly representing a basic type—are not the most characteristic for the class. It is the Liliaceous floral scheme, generally indicated by the expression, $P\,3+3$, $A\,3+3$, $G\,3$, which is

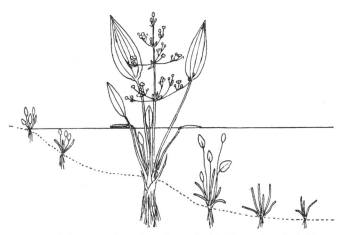

Fig. cxli. Diagrammatic section of a pond with Water-plantain (*Alisma Plantago*, L.), showing seedlings in deeper water with ribbon-leaves only, and in shallow water with aerial leaves; also a full-grown flowering plant with submerged, floating and aerial leaves. [Lister, G. (1920).]

recurrent throughout the Monocotyledons. I have written "$G\,3$," because this is the accepted Liliaceous formula, but I think that the recent work of E. R. Saunders[2], which we shall discuss later (pp. 197—199), makes it probable that it would be more correct to write "$G\,3+3$," and thus to suppose that the flowers are constructed on a hexacyclic, trimerous plan. This Liliaceous flower-type extends throughout the Liliiflorae—the group of families of which the best known are the Liliaceae, Iridaceae and Amaryllidaceae—and it shows itself susceptible of an amazing range of variation. The perianth may either consist of six similar members (as in most of the flowers shown in fig. cxliii, p. 186) or there may be a differentiation between the inner and outer members, as in the species of *Iris* drawn in fig. I A, where

---

[1] Arber, A. (1920[4]), pp. 9—23, etc.   [2] Saunders, E. R. (1923).

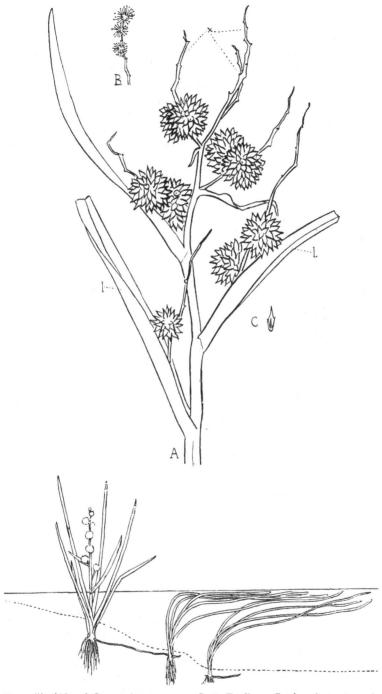

Fig. cxlii. (Above) *Sparganium ramosum*, Curt., Fordham, Cambs., Aug. 17. A, in-fructescence (× ½); leaves, *l.* and *l.*, cut short ; the bare branches from which the male heads have fallen are marked with a × ; B, small branch of staminal part of younger inflorescence ; C, single nutlet with perianth. [A.A.] (Below) *Sparganium simplex*, Huds.; deep and shallow water plants connected by runners; the stream bed is seen in section. [Lister, G. (1920).]

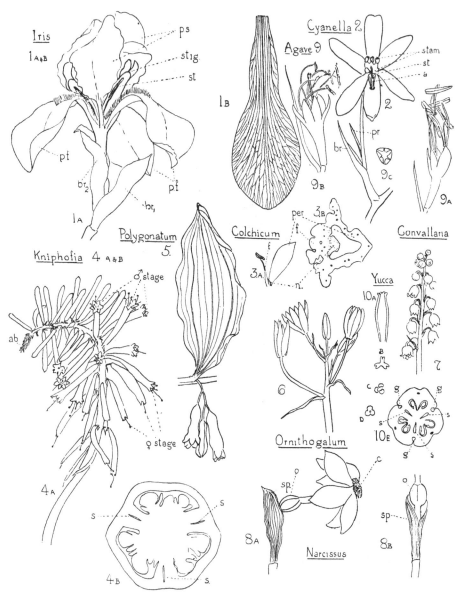

Fig. cxliii. Some flower types of Liliiflorae. Figs. 1 A and B, *Iris* (Iridaceae); fig. 1 A, flower of *Iris* sp. ($\times\frac{1}{2}$); *br₁* and *br₂*, bracts; *p.f.*, three bearded outer perianth members; *p.s.*, erect inner perianth members; *st.*, stamen; *stig.*, one of three bifid petaloid stigmas; fig. 1 B, perianth segment of *Iris hexagona*, Walt. ($\times\frac{1}{2}$), to show venation, which is slightly simplified (Brit. Mus. Herbarium). Fig. 2, *Cyanella* sp. (Amaryllidaceae); *br.*, bract; *pr.*, prophyll (bracteole); *stam.*, five staminodes; *a.*, single fertile stamen; *st.*, style. Figs. 3 A and B, *Colchicum* sp. (Liliaceae); fig. 3 A, stamen, and perianth segment to which it is attached ($\times\frac{1}{2}$); fig. 3 B, transverse section just below attachment of filament to perianth ($\times$ 14); both figs. show nectar-secreting swelling, *n.* (dotted), just above attachment of filament, *f.*, to perianth member, *per.* Figs. 4 A and B, *Kniphofia rufa*, Hort. (Liliaceae); fig. 4 A, inflorescence ($\times\frac{1}{2}$); *ab.*, abortive flowers; fig. 4 B, transverse section of young ovary ($\times$ 23) to show position of septal glands, *s*. Fig. 5, *Polygonatum* sp., leaf with axillary group of flowers ($\times\frac{1}{2}$). Fig. 6, *Ornithogalum* sp., inflorescence ($\times\frac{1}{2}$). Fig. 7, *Convallaria majalis*, L., inflorescence ($\times\frac{1}{2}$). Figs. 8 A and B, *Narcissus* sp., flower with ovary, *o.*, and corona, *c.*, enclosed in spathe, *sp.*, ($\times\frac{1}{2}$). Figs. 9 A—C, *Agave dasylirioides*, Jacobi et Bouché; fig. 9 A, flower at ♂ stage ($\times\frac{1}{2}$); fig. 9 B, older flower at ♀ stage ($\times\frac{1}{2}$); fig. 9 C, top of stigma, enlarged. Fig. 10, *Yucca* sp.; fig. 10 A, side view of ovary ($\times\frac{1}{2}$); figs. 10 B—D, relation of stigmatic branches and ovary; fig. 10 E, transverse section of ovary; *g.*, honey grooves; *s.*, septal glands. [A.A.]

there is a marked distinction between the outer bearded "falls" and the inner erect "standards"; in some cases the differentiation may lead to zygomorphy, e.g. *Gilliesia*. The members of the perianth, again, may be free, as in *Ornitho-galum* (fig. 6), or united, as in *Kniphofia, Polygonatum* and *Convallaria* (figs. 4, 5, 7). The intercalary growth of the basal region of the united perianth members sometimes results in the development of a surprisingly long flower tube. Those in the flowers of *Colchicum illyricum*, Stokes, drawn in fig. xxiii, 1 A, p. 46, were respectively 16 cms. and 19 cms. long; and in the case of two Amaryllids—*Cooperia Drummondii*, Herb., and *Pancratium trianthum*, Herb.—I have measured perianth tubes fully 12 cms. long. In all these plants, and, indeed, in most of the Liliiflorae, the perianth is conspicuous and often beautifully coloured; but there are also instances of insignificant flowers, which may even fail to open. *Juncus bufonius*, L., the Toad-rush, for instance, is always cleistogamic. In this species, the tube formed by the perianth members, which touch above the style, acts as a barrier to the emergence of the three stigmatic branches. They are obliged to recurve towards the base, and their papillae come into contact with the little anthers at the sides of the ovaries. The pollen passes by a terminal pore, so fertilisation takes place very straightforwardly[1].

The stamens of the Liliiflorae, though they are commonly six in number, may be reduced to three, as in *Iris*, or certain of them may be developed as staminodes; fig. cxliii, 2, shows the flower of *Cyanella* (Amaryllidaceae) in which there are five staminodes and one fertile stamen. On the other hand, the number of stamens may be increased. In *Vellozia* there are six stamen fascicles—each consisting of three to ten stamens—instead of six single stamens. The corona, which in certain Amaryllids, such as *Narcissus Bulbocodium*, L., the Hoop-petticoat Daffodil, exceeds the perianth in size and importance, is generally recognised as equivalent to fused stipular lobes belonging to the stamens.

The gynaeceum in this cohort, like the other parts of the flower, is remarkably variable, both in its general relations and in its structure. It may be superior, as in *Kniphofia*, etc. (fig. 4 A), or inferior, as in *Narcissus* (fig. 8). The style may be simple, as in *Cyanella* and *Agave* (figs. 2 and 9), or with three branches, as in *Colchicum*. The stigmas in *Iris* are located on three petaloid stylar lobes, while in *Aspidistra* the style spreads out into a mushroom-like head, which bears the receptive surfaces.

So far we have mentioned only hermaphrodite flowers, but within the Liliiflorae we also find examples of unisexuality, e.g. Black Bryony, *Tamus communis*, L. (frontispiece). Here the flowers are greenish and inconspicuous, but the male inflorescences exhale a strong honey scent.

[1] Laurent, M. (1904).

Apart from the variety in the individual flowers, the reproductive shoot system, as a whole, varies greatly in complexity; there is every range from a flower borne singly on a solitary peduncle, to a gigantic, many-flowered inflorescence. But this catalogue of the variation in the reproductive shoots of the Liliiflorae might be continued almost endlessly—enough has, perhaps, been said to show that the Liliaceous type of flower is capable of a remarkable degree of adjustment within the limits of its definite, but elastic, basal scheme. An equally striking series of variations might be traced within the Farinosae, a cohort which seems to show affinity with the Liliiflorae. *Pontederia* (Pontederiaceae) is drawn in fig. cxxxviii, 3, p. 181, and, in fig.

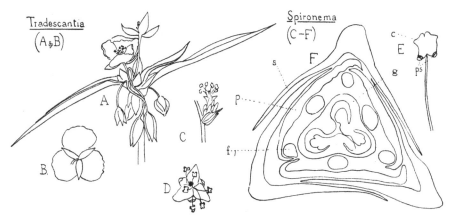

Fig. cxliv. Flowers of two Commelinaceae. A and B, *Tradescantia virginiana*, L.; A, inflorescence (× ½); B, flower viewed from below (× ½); C—F, *Spironema fragrans*, Lindl.; C, inflorescence with chaffy bracts (× ½); D, flower from above; E, single stamen, magnified to show exaggerated connective, *c.*, and small pollen-sacs, *p.s.*; F, transverse section from microtome series through bud (× 14); *s.*, sepals; *p.*, petals; *f.*, filaments; *g.*, gynaeceum. [A.A.]

cxliv, two flowers are shown from the Commelinaceae, another of the families of this group. Fig. cxliv, F, represents a transverse section from a microtome series through a bud of *Spironema fragrans*, Lindl. I give this sketch because I feel that the average floral diagram is, in many ways, an unsatisfactory combination of theory and observation; I think that it would make for exactness of thought, if it could be supplemented, wherever possible, by sections from transverse microtome series through buds, in which the *actual* relations of the parts are revealed. As a contrast to the Liliiflorae and Farinosae, we may take an instance from the Araceae. The variation of flower structure within this family is amazing, and it is impossible to deal with it here; we shall consider it in general terms in the next chapter (pp. 212—216). But we may now choose our native

Arum maculatum, L., as an example—not as a *type*, since it would be absurd
to pretend that any one form can be typical for such a range of contrasting
reproductive shoots. The general construction of the inflorescence of "Lords
and Ladies" is thoroughly familiar (figs. cxlv, 2 A and D, p. 190); the spadix,
enclosed in the spathe, first bears female flowers (♀), then a few sterile (*st.*),
then male (♂), and, finally, sterile again (*st.*), while the top of the club (*cl.*)
is generally described in the text-books as a naked axis. In examining the
inflorescence I was much intrigued by this club, because its leafless character
would be, as Church noticed in 1908, "a most anomalous condition[1]." Since
these words were written, the conception of the club as a leafless axis has
become more, instead of less, puzzling, since the leaf-skin theory[2] renders it
highly unlikely that any genuinely naked axis exists among the Flowering
Plants. With this in mind, I cut the whole spadix lengthways (fig. cxlv, 2 E),
and I was interested to find that it is only the core of the club which looks like
a continuation of the axial part of the spadix; the outer region is yellowish,
and resembles, in colour and consistency, the fused bases of the male flowers,
while the maroon velvety surface recalls the hue and texture of the anthers.
So it occurred to me that possibly the club represents a region of the
inflorescence in which the sterilisation of the flowers has been carried to a
farther point than in those marked *st.*, so that they remain rudimentary
and unindividualised. Since these hypothetical flowers are completely fused
with one another, and with the stem that bears them, they merely form a
superficial layer, like a thick elongated thimble, clothing the apical region
of the inflorescence axis. For this notion, though it seemed to me probable,
I could, however, find no proof; but on referring to Engler's work on the
Araceae[3], I learned that he had not only anticipated me in this idea, but
that he had given convincing proof of its validity. He shows, by comparative
study, that the club is not a naked axis, but that it consists of an incompletely developed part of the inflorescence. *Arum*, on his view, forms the
end term of a series beginning with *Helicodiceros muscivorus*, (L.) Engl., in
which the club is clothed to the apex with awl- or bristle-shaped flower
rudiments; this series continues through *Helicophyllum crassifolium*, (Ledeb.)
Engl., in which the flower rudiments, though recognisable as such, are
reduced to mere warts; and finally to *Arum*, in which the peripheral tissue
of the club corresponds to the tissue of stamens whose individuality has
been lost.

The male flowers are said to consist of paired stamens without a perianth,
and one can sometimes see evidence of this, but the general effect is that of
a crowd of 4-locular anthers (fig. 2 F), showing no connexion with one another. The female flowers consist of single gynaecea, each of which is described

[1] Church, A. H. (1908). [2] Saunders, E. R. (1922).
[3] Engler, A. (1881—84); see Bd. v, 1884, pp. 298—9.

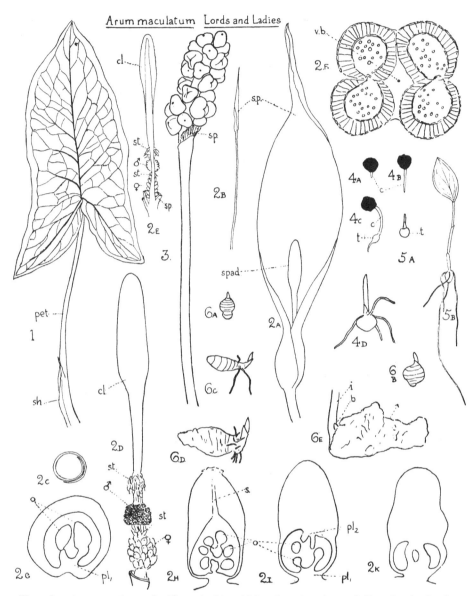

Arum maculatum Lords and Ladies

Fig. cxlv. *Arum maculatum*, L. Fig. 1, leaf in which only main veins are indicated; *sh.*, sheath; *pet.*, petiole; *l.*, limb (×½). Figs. 2 A—K, inflorescence; fig. 2 A, spathe, *sp.*, with spadix, *spad.* (×½); fig. 2 B, inflorescence with peduncle complete (×¹⁄₁₀); fig. 2 C, section at base of spathe showing "wrap-over"; fig. 2 D, spadix, with spathe cut away, to show ♀, female flowers; *st.*, sterile flowers; ♂, male flowers; *cl.*, club (×1); fig. 2 E, spadix cut longitudinally, lettered as in fig. 2 D (×½); fig. 2 F, transverse section of an anther (×23); *v.b.*, vascular bundle; fig. 2 G, transverse section of an ovary (×14); *o.*, ovules; *pl₁*, placenta; figs. 2 H—K, transverse sections from below upwards from microtome series through another ovary in which a second placenta, *pl₂*, with two ranks of ovules is developed. Sections are transverse to spadix, and thus slightly oblique to ovary; *s.*, stylar canal (×14). Fig. 3, spadix with berries (×½), Chiddingly, Sept. 1, 1922; *sp.*, shrivelled spathe. Figs. 4 A—D, stages in germination of cultivated seedlings; seeds sown Sept. 14, 1922; *c.*, cotyledon; *t.*, tuber; figs. 4 A—C, stages reached on April 21, 1923 (×1); fig. 4 D, stage reached Oct. 9, 1923 (×1). Fig. 5 A, a seedling found at Chiddingly, Sussex, Sept. 7, 1922, probably at same age as fig. 4 D (×½); fig. 5 B, the seedling drawn in fig. 5 A, at the stage which it had reached on Feb. 11, 1923 (×½). Figs. 6 A—E, tubers of various ages found on digging about in neighbourhood of fruiting spikes, Chiddingly, Sept. 7, 1922 (×½); fig. 6 E, *i.*, base of infructescence; *b.*, bud; the tuber, which was covered with dead scale-leaves and roots, was hollow and empty behind the point marked with an arrow. [A.A.]

as unilocular, with ovules on its posterior margin alone (fig. 2 G). But it is, I think, of interest—as showing that there is still plenty of room for morphological work, even on our commonest native plants—that, in cutting serial sections through the female part of an *Arum maculatum* spadix, chosen at random, I found evidence of dimery in the ovary; the ovules were borne on two placentae, adaxial and abaxial (figs. 2 H—K). It is, however, by no means surprising that such a type of structure should occur in this genus, for Engler's work has made it clear that the Aroids show various reduction stages from a tri-carpellary ovary. The infructescence (fig. 3) consists of a spike of soft scarlet berries, irregular in form, sessile, but very readily detachable, and leaving a vertical, elongated, orange scar on the thick pale spadix. There are generally two large seeds, but in those I have examined the number varied from one to four.

To show how different other Araceae may be from the *Arum* type, I have illustrated in fig. cxxxviii, 6 A—C, p. 181, the inflorescence and flowers of *Spathicarpa sagittifolia*, Schott. Here the spadix is completely fused to the spathe; the anthers are grouped in synandria, *s.*, raised on minute columns, while the female flower consists of a four-membered perianth and a gynaeceum. Another widely different type is that of the Skunk-cabbage of N. America, *Symplocarpus foetidus*, Nutt., with its hermaphrodite flowers (figs. cxlvi, C and D, p. 192).

In considering *Arum maculatum*, I have not described the mode of cross-pollination, because it is so well known, and an account of it finds a place in every text-book. It seems to me, indeed, that the subject of the agencies by which pollen is transferred to the stigma in the Flowering Plants has attracted an altogether disproportionate amount of attention; the fact is that anything in which nature displays—or seems to display—the quality which in human beings would be called ingenuity, is at once unduly stressed, because it appeals to the anthropocentric view of the universe, which none of us can completely escape. So, since I feel that they habitually receive too much of the limelight, I propose to pass lightly over the pollination mechanisms of Monocotyledons, which show at least as great a range as those of Dicotyledons; wind, water, insects and birds, in turn prove to be the agents by which the pollen is carried.

Wind-pollination is characteristic of many Monocotyledons. The Palms come to one's mind at once as the classic instance of anemophily, because, in the regions which have been the cradle of western civilisation, the Date-palm has been of paramount importance to man from time immemorial. The female tree's need of pollen was recognised in early days, and the too-erratic action of the wind was often replaced by human agency, in order to ensure a successful crop. Among the Assyrian bas-reliefs from Nimrûd, dating from the ninth century before Christ, which are now in the British Museum, there are wonderful winged figures associated with conventionalised "sacred trees," each of which probably represents a male Palm growing in the midst of a grove of female trees, festooned together for the annual festival of fertilisation. In such an example as that shown in fig. cxlvii, the basket held in the left hand of the Winged Being is supposed to correspond

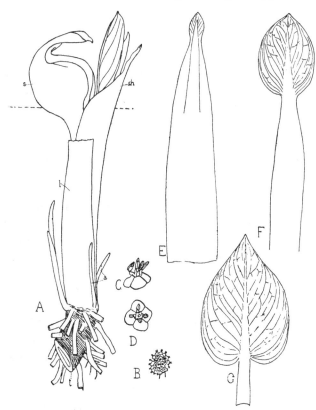

Fig. cxlvi. *Symplocarpus foetidus*, Nutt., Skunk-cabbage [A.A.]. Cambridge Botanic Garden, March. A, plant (× ⅛); dotted line, ground level; *s.*, spathe, dark purple, almost black; *sh.*, leafy shoot; *l.*, leaf of which the upper part has perished; *t.*, tuberous axis, with adventitious roots which are cut off short; *a.*, axillary buds; B, spadix (× ⅓); C, flower from the side (enlarged); D, flower from above (enlarged); E, F, G, leaves from shoot (× ⅓); E is apparently the second leaf, which follows a similar first leaf; F, third leaf; G, sixth leaf; this leaf, which is still immature, shows the relatively great size gradually attained by the limb. [A.A.]

to the receptacle which is, even to-day, carried by the Arab who climbs the male tree to secure the stamen clusters, and the fir-cone-like object in the other hand is the male inflorescence itself: the bird-headed figure is thus undoubtedly in the act of pollinating the Date-palm[1]. We are told that Assyriologists have no certain knowledge of the nature and attributes of

[1] Tylor, E. B. (1890) and Garlick, C. (1918); the latter writer approaches the question from a more definitely botanical standpoint.

these mythical winged creatures, so the botanist may perhaps be forgiven for adopting them as the tutelary gods of anemophily.

In addition to the Palms, we have another great wind-pollinated group in the Grasses. Here we find anemophily associated with a simplified flower, in which there is nothing that can be called a typical perianth, although the little structures known as lodicules (*l.*, fig. cxlviii, A, p. 194) are sometimes classed as perianth members.

Fig. cxlvii. Eagle-headed Winged Being, in the character of fertilizer of the Date-palm. After a bas-relief from the Palace of Ashur-nasir-pal at Nimrûd. (British Museum.)

Aquatic pollination, though unusual, is less infrequent in Monocotyledons than in Dicotyledons, but I will not attempt to discuss it here, since to do so would involve a repetition of what I have said in a former book[1].

As among Dicotyledons, pollination by members of the animal kingdom is associated with excess pollen or honey, which serves as food to the visitor. In *Antholyza* (Iridaceae), bird-pollination of the scentless, strongly zygomorphic, protandrous flowers has been observed, but such ornithophily is a rarity among Monocotyledons[2].

Insect-pollination as a rule follows the same general lines as in Dicotyledons. The inflorescence of *Kniphofia rufa*, Hort., shown in fig. cxliii, 4 A, p. 186,

[1] Arber, A. (1920⁴), pp. 235—8, etc.   [2] Porsch, O. (1910).

is a typical example of protandry. The flowers at the tip of the shoot are abortive; next come unopened buds; these are followed by open flowers, whose stamens soon dehisce, but in which the stigma has not emerged; and finally we reach flowers at the female stage, whose stamens have shrivelled, while the stigma is receptive and is tipped by a clear sticky drop of fluid. A similar protandry will be recognised in the two flower stages of *Agave* drawn in figs. cxliii, 9 A and B, while in other genera the converse case of protogyny occurs.

Nectar secretion is distributed in Monocotyledons as variously as in Dicotyledons. In *Colchicum* the outer side of the filament-base secretes honey, which collects in a small pit formed by the junction of filament and perianth (figs. cxliii, 3 A and B, p. 186). Many flowers secrete honey from

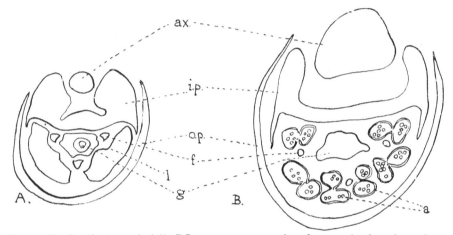

Fig. cxlviii. *Danthonia provincialis*, DC., two transverse sections from a series through one flower (× 46); A, lower; B, higher; *ax.*, axis; *o.p.*, outer palea; *i.p.*, inner palea; *l.*, lodicules; *f.*, filaments; *a.*, anthers; *g.*, gynaeceum. [A.A.]

the perianth (e.g. *Uvularia*, fig. xxxv, 3 B, p. 58), while in others the nectaries are formed in the septa of the gynaeceum. These septal nectaries[1] may exist alone, as in *Kniphofia* (fig. cxliii, 4 B), or there may also be external honey-secreting grooves on the same radius of the ovary, as in *Yucca* (fig. cxliii, 10 E).

It is impossible here even to touch upon the manifold complexities associated with the insect-pollination in such groups as the Scitamineae, the tropical members of the Orchidaceae, etc. I have shown the flower structure of *Roscoea* in fig. ix, p. 22, and, as regards the Orchids, I shall confine myself to illustrating some of the types met with in the British flora. A detailed account of the structure of these plants, and observations on their pollination, are given in Charles Darwin's book on the fertilization of

[1] Grassman, P. (1884); Saunders, E. R. (1890); Schniewind-Thies, J. (1897).

Orchids[1], but this work has, unfortunately, never been produced in a fully illustrated edition, so the sketches given in figs. cxlix and cl may perhaps be of a little service in supplementing Darwin's figures. The general relations of the parts of the flower in the Orchids and the manner in which cross-fertilization comes about, are so familiar to students that I will only here refer to one or two anomalous cases, which seem to be of some theoretical interest.

In fig. cxlix I have illustrated the structure of *Cephalanthera pallens*, Rich. (described by Darwin under the name of *C. grandiflora*). The flower is peculiar in having no rostellum, and in the fact that the pollen forms friable pillars instead of firm pollinia. Darwin observed that, from these pollen pillars, a multitude of pollen tubes penetrated the stigmatic tissue; and it appears from his account that self-fertilization is a common occurrence, and that the plant thus offers a partial exception to the rule that the flowers of Orchids are generally fertilized by pollen from another plant. It must therefore be admitted that in this species the general formation of the flower seems to serve little purpose. A still more striking instance of useless "adaptation" is offered by the Bee-orchis, *Ophrys apifera*, Huds. This flower in its construction differs in no essential respect from the general Ophrydean type; its appearance is shown in fig. cl, I A, while the anther, *an.*, the paired pollen masses, *p.*, and the paired rostella, *r.*, are drawn on a larger scale in fig. I B. The stalks of the pollinia, however, differ from those of the rest of the tribe in being extremely flexible, with the result that, when the anthers open, the pollen masses drop forward and hang directly opposite the stigma, so that the slightest breath of wind swings them against the receptive surface to which they adhere. Darwin records the results

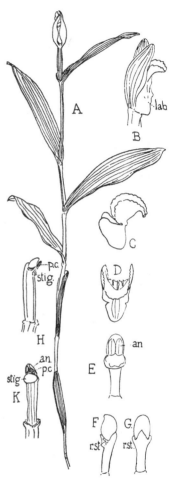

Fig. cxlix. *Cephalanthera pallens*, Rich. (*C. grandiflora*, S. F. Gray). A, flowering shoot (× ⅓). B, side view of flower; labellum, *lab.*, in natural position. C and D, side and front views of labellum, upper part bent down to show hinge action. E—G, column from unopened bud, anthers, *an.*, unopened; E, front; F, side; G, back; *r.st.*, rudimentary stamen. H and K, column from fading flower; *p.c.*, friable pillars of pollen, which are dropping forward unnaturally; *stig.*, stigma; *an.*, anther. [A.A.]

[1] Darwin, C. (1904).

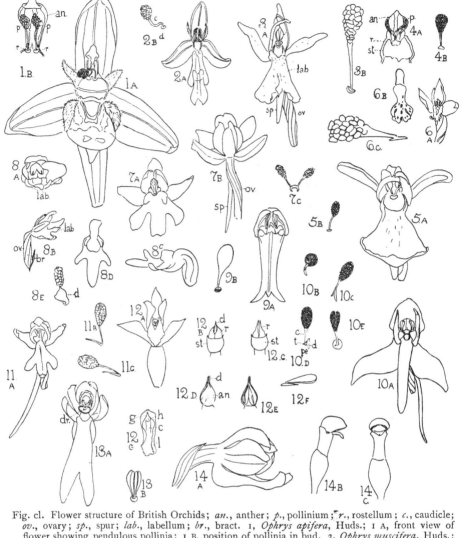

Fig. cl. Flower structure of British Orchids; *an.*, anther; *p.*, pollinium; *r.*, rostellum; *c.*, caudicle; *ov.*, ovary; *sp.*, spur; *lab.*, labellum; *br.*, bract. 1, *Ophrys apifera*, Huds.; 1 A, front view of flower showing pendulous pollinia; 1 B, position of pollinia in bud. 2, *Ophrys muscifera*, Huds.; 2 A, front view of flower; 2 B, a pollinium removed and viewed as from side of flower. 3, *Orchis maculata*, L.; 3 A, front view of flower; 3 B, pollinium. 4, *Orchis mascula*, L.; 4 A, centre of flower; 4 B, pollinium. 5, *Orchis latifolia*, L.; 5 A, front view of flower; 5 B, pollinium. 6, *Orchis ustulata*, L. (Swiss specimen); 6 A, old flower; 6 B, labellum of young flower; 6 C, pollinium. 7, *Orchis pyramidalis*, L.; 7 A, front view of flower; 7 B, back view of flower; 7 C, pollinia. 8, *Herminium Monorchis*, R.Br.; 8 A, view looking obliquely into flower; 8 B, side view of flower from below; 8 C, labellum and column from the side; 8 D, labellum and column from above; 8 E, pollinium with helmet-like disc. 9, *Habenaria viridis*, R.Br. (*Peristylus viridis*, Lindl.); 9 A, front view of flower; 9 B, pollinium. 10, *Habenaria chlorantha*, Bab.; 10 A, front view of flower; 10 B—D, successive drawings of pollinium removed from flower, in side view; *t.*, rudimentary tail of caudicle; *pe.*, drum-like pedicel; 10 E, pollinium, showing disc in surface view. 11, *Habenaria conopsea*, Benth.; 11 A, front view of flower; 11 B and C, a pollinium showing change of position on removal from flower; the pollinium bends forward until its caudicle is almost parallel with the surface of the disc. 12, *Spiranthes autumnalis*, Rich.; 12 A, front view of flower; 12 B, central part of flower, perianth removed, viewed from rostellum side; the disc, *d.*, is seen through the rostellum membrane; 12 C, same, after stroking lower side of rostellum with needle, and so removing disc and pollen-masses; 12 D, in same state as 12 B, but viewed from opposite side; the disc, *d.*, has the rostellum membrane folded over it; 12 E, same as 12 D, but with anther removed; 12 F, pollen-masses and disc removed from flower; 12 G, labellum with *h.*, hollow for nectar; *c.*, central channel; *g.*, globular nectar-secreting process; *l.*, reflexed and fringed lip. 13, *Listera ovata*, R.Br.; 13 A, front view; *dr.*, drop of viscid matter which has extruded explosively; 13 B, pollinia. 14, *Neottia Nidus-avis*, Rich.; 14 A, side view of flower; 14 B and C, side and front views of ovary and column from unopened flower. [A.A.]

of many field observations upon the Bee-orchis, from which it transpires that self-pollination is the rule. As he points out, the general construction of the flower is of the type "adapted" to cross pollination, but this complex mechanism is useless, since the plant pollinates itself. Darwin adds, "The whole case is perplexing in an unparalleled degree, for we have in the same flower elaborate contrivances for directly opposed objects." It must, indeed, be admitted that the consideration of this flower opens up a series of difficult questions, but the perplexity aroused by it can only be described as "unparalleled," if we are determined to explain everything as the result of adaptation. A drawback of the teleological standpoint is that it so often leads to tilting at windmills—encouraging elaborate attacks on problems which have no existence, except in the minds of those to whom all structure is directly purposeful.

There is another case of self-fertilization among the Orchidaceae, which is of a more sensational type than the two already mentioned. The Bird's-nest Orchid, *Neottia Nidus-avis*, Rich., frequently forms flowering axes, which instead of rising vertically into the air, show a growth-curvature which prevents their reaching the surface of the soil. These subterranean inflorescences are self-fertilized in the humus, and the seeds, which have no opportunity of escaping, germinate where they are formed[1].

Apart from their relation to insects, the flowers of the Orchidaceae show various features of structural interest, to one of which we must now refer. The gynaeceum in this family has either one or three loculi—the change from one to three sometimes occurring within an individual ovary, e.g. *Apostasia nuda*, R.Br. (figs. cli, D and E, p. 198). Until recently it has been regarded as tricarpellary—the carpels being considered to be fused by their ovule-bearing edges, while the stigmas were interpreted as standing over the carpellary midrib (fig. cli, A). But in a paper published in 1923, E. R. Saunders has suggested a different view, which makes the gynaeceum hexamerous. She regards the carpels as of the "semi-solid" type, the ovule-bearing margins being carried inwards until they come to stand to right and left of the midribs, that is to say, of those veins which, on the older view, occur at the carpellary margins. The veins, on the other hand, which are continued up towards the stigmas, and which, from the orthodox standpoint, are treated as carpellary midribs, are interpreted by Saunders as belonging to a second whorl of sterile solid carpels. The difference between the two views is perhaps a little difficult to grasp, when it is expressed in words, but diagram A, in fig. cli, will, it is hoped, make it clear. The new view has the advantage of explaining the mode of dehiscence of certain Orchid ovaries, which split into three wide seed-bearing valves, and three narrow strip-like

[1] Bernard, N. (1902).

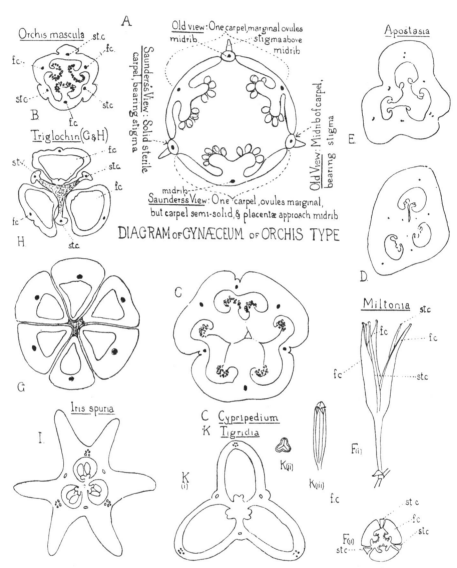

Fig. cli. Sketches illustrating the gynaeceum theory of E. R. Saunders (1923): throughout, *f.c.*, fertile carpel; *st.c.*, sterile carpel. A, diagram contrasting the orthodox view of the Orchid gynaeceum with that of E. R. Saunders. B, transverse section of ovary of *Orchis mascula*, L. (× 14). C, transverse section of ovary of *Cypripedium* sp. (× 14). D and E, transverse sections of ovary of *Apostasia nuda*, R.Br. (× 23); D, near base; E, higher up; in the herbarium material used the small delicate ovules were poorly preserved. F, *Miltonia vexillaria*, Nichols, var. "Constance"; F (i), capsule after dehiscence (× ½); F (ii), transverse section of dehiscing ovary; the seeds had been shed in the ovary examined, but their probable position is indicated in the diagram. G, *Triglochin maritima*, L., transverse section of ovary (× 14). H, *Triglochin palustris*, L., transverse section of ovary (× 23). I, *Iris spuria*, L., transverse section of ovary (× 4½); small lacuna in centre of section. K, *Tigridia Pavonia*, Ker-Gawl.; K (i), transverse section of ovary, enlarged, ovules omitted; K (ii) and K (iii), ovary from the top and side (× ½). [A.A.]

sterile valves (e.g. *Miltonia*, fig. cli, F). The gynaeceum of the Iridaceae (figs. cli, I and K) may perhaps be explained on the same hypothesis[1]. It may, of course, be objected that in all these cases the ovaries concerned are inferior, and hence that the construction is complicated by the ribs and strands which are related to the perianth and androecium. I have considered the Orchids in this connexion, because they were the first Monocotyledonous family to be discussed by Saunders, but there is little doubt that the theory is equally applicable to various members of the class in which the flower is hypogynous. That it is possible for a superior whorl of fertile carpels to be replaced by a whorl of solid sterile carpels is proved by the comparison of the gynaeceum of the Sea-arrowgrass, *Triglochin maritima*, L., with that of the related species, *T. palustris*, L. (figs. cli, G and H). These gynaecea are of a different type, however, from those of the Orchids, since the carpels are follicular.

When we turn from the general structure of the gynaeceum to the detailed examination of the ovules, we find that, since modern methods of microtechnique were evolved, we have amassed an immense store of information about the minute structure and development of the gametophytes, the process of fertilization, and the early history of the embryo, in the Flowering Plants. Monocotyledons have been studied even more fully than Dicotyledons, since they often offer peculiarly favourable material; the large nuclei of the Liliaceae, for instance, make it relatively easy to follow the events that lead to the production of the embryo. American workers have been particularly active in this field, and in Coulter and Chamberlain's *Morphology of Angiosperms* we have a book in which Monocotyledons and Dicotyledons are treated, primarily from this standpoint, by professors of the Chicago school. This work, with the references which it includes, furnishes a clue to the research that has been done in this direction, and I shall not attempt to summarise it here, since the results are so voluminous as to defy brief analysis. But since we have been particularly concerned with the flowers of the Liliiflorae, I add here a sketch showing the actual process of "double fertilization" in *Lilium Martagon*, L. (fig. clii, A, p. 200), while fig. clii, B, shows a young embryo in an embryo-sac, of which the small antipodal end alone remains unchanged, thus affording a standard by which we can realise the alteration in size of the sac since fertilization. And as an example of the development of integument and embryo in the fertilized ovule of another of the Liliiflorae, I also include drawings of *Juncus bufonius*, L., from Laurent's beautiful memoir on the Juncaceae (fig. cliii, p. 201). In the oldest stage shown (fig. cliii, C), the great mass of the embryo is formed from the cotyledon, which enwraps the small plumular rudiment, *pl.* The hypocotyl

---

[1] Since this book was in print, a further instalment of Saunders's work has appeared: "On Carpel Polymorphism. I," *Ann. Bot.*, vol. XXXIX, 1925, p. 123.

is not yet recognisable; the piliferous layer and root-cap originate from persistent suspensor cells.

The fruits of the Liliiflorae are—like the flowers—of varied types. Fig. cliv, p. 202, shows examples of two contrasting forms—the loculicidal capsule (*Fritillaria Imperialis*, L., and *Iris foetidissima*, L.) and the berry (*Convallaria majalis*, L.). The raceme of orange-coloured fruits of the Lily-of-the-valley is a very striking object; the perianth, reduced to papery consistency, persists, as does the style.

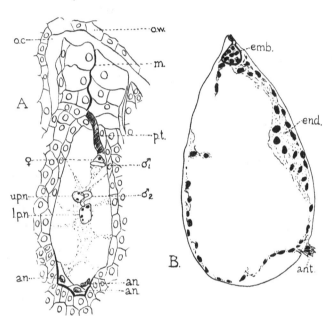

Fig. clii. *Lilium Martagon*, L. A, longitudinal section of embryo-sac to show double fertilization (× 180); *o.w.*, ovary wall; *o.c.*, ovary cavity; *m.*, micropyle; *p.t.*, pollen tube; ♀ egg-cell nucleus; ♂₁ and ♂₂, the two male nuclei; *u.p.n.*, upper polar nucleus; *l.p.n.*, lower polar nucleus; *an.*, three antipodal nuclei. B, longitudinal hand section through an embryo from a flower on the sixteenth day of opening; *emb.*, small embryo at upper end of sac; *end.*, endosperm lining sac (torn in cutting the section); *ant.*, antipodal end of sac, which does not keep pace in growth with the rest (× 75). [A.A.]

The seeds of nearly all Monocotyledons, as of Dicotyledons, are shed, and germinate apart from the parent plant. True vivipary occurs, however, as a rarity among the Grasses (e.g. *Melocanna*[1]), and in one of the marine Potamogetonaceae — *Pectinella antarctica*, Black (*Cymodocea antarctica*, Endl.). *Pectinella* was mentioned in my book on Water Plants, but was incompletely discussed[2] since I was then unacquainted with the work of J. M. Black[3], whose illustrations I am allowed to reproduce in fig. clv, p. 203.

[1] Stapf, O. (1904).   [2] Arber, A. (1920⁴), p. 127.   [3] Black, J. M. (1913).

Black has definitely settled the controversial question of the nature of the cup-like body, or "comb," which the seedling carries with it, when it finally breaks away from the parent (figs. clv, 7 and 14). It seems that this cup is the horny four-lobed frame-work of the pericarp. As soon as fertilization is accomplished, the long stigmas break off and the embryo begins to grow.

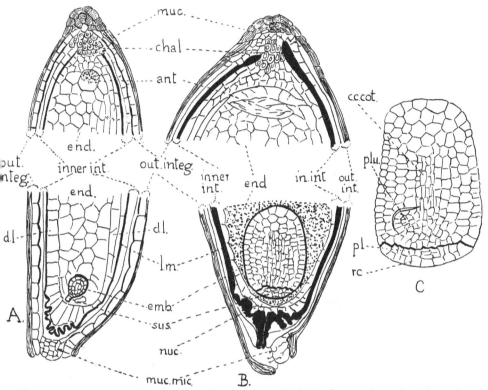

Fig. cliii. Development of ovule and embryo in *Juncus bufonius*, L. A and B, micropylar and chalazal ends of young and older ovule (× 225); *muc.*, mucilaginous outer layer of outer integument; *l.m.*, lignified membrane of inner integument; *chal.*, chalaza; *d.l.*, digestive layer; *end.*, endosperm, which in B is partly digested; *ant.*, tissue derived from antipodals; *emb.*, embryo; *sus.*, persistent suspensor tissue; *nuc.*, remains of nucellar tissue; *muc. mic.*, mucilaginous tissue blocking micropyle. C, embryo almost mature (× 270); *p.l.*, piliferous layer and *r.c.*, root-cap, both formed from persistent suspensor tissue; *plu.*, plumule; *c.c.cot.*, central cylinder of cotyledon; hypocotyl not yet differentiated. [Adapted from Laurent, M. (1904).]

The base of the style splits, and the plumule emerges. By the decay of the softer tissues the cup is reduced to a four-lobed comb, which anchors the seedling when it is eventually detached. The viviparous seedling floats, resembling in this respect the fruit of *Posidonia australis*, J. D. Hook, which is buoyed up by aerenchyma in the exocarp. These two plants thus form an exception to the rule that the marine Monocotyledons do not possess any special adaptation for seed or fruit dispersal. But as Ostenfeld[4] has pointed

[1] Ostenfeld, C. H. (1916).

out, they nevertheless occupy unusually restricted areas; this fact seems to harmonise with Willis's contention that it is the age of a species, rather than its "adaptations" for dispersal, which determines the area over which it is distributed[1].

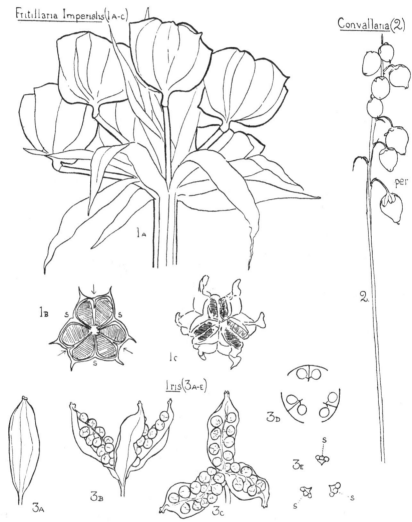

Fig. cliv. Fruits of some Liliiflorae. Figs. 1 A—C, *Fritillaria Imperialis*, L.; fig. 1 A, infructescence, July 1 (× ½); fig. 1 B, transverse section of fruit (× ½); seeds shaded; arrows indicate future lines of dehiscence; *s.*, septum; fig. 1 C, dehiscing fruit viewed from top, September 14; seeds shaded (× ½). Fig. 2, *Convallaria majalis*, L., infructescence, August 29; *per.*, perianth (× ½). Figs. 3 A—E, *Iris foetidissima*, L.; fig. 3 A, undehisced fruit (× ½); figs. 3 B and C, dehisced fruits (× ½); fig. 3 D, diagrammatic section of three valves of fruit after dehiscence; fig. 3 E, diagram of top of three valves of dehisced fruit; *s.*, top of septum. [A.A.]

[1] Willis, J. C. (1922).

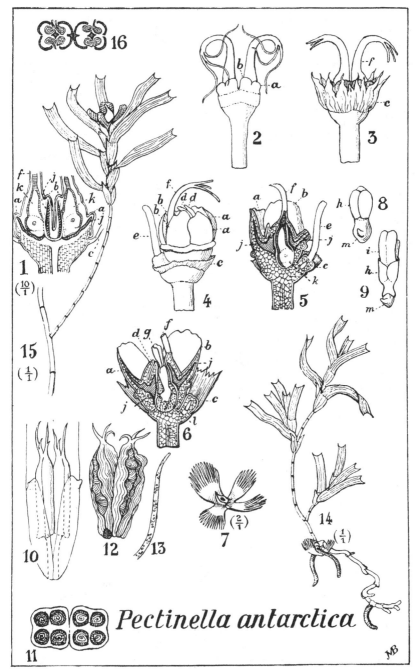

Fig. clv. *Cymodocea antarctica*, Endl. = *Pectinella antarctica*, Black. Fig. 1, vertical section of female flower at about the same stage as fig. 2. Fig. 2, young female flower (from long-leaved variety). Fig. 3, female flower after fertilization (from short-leaved form). Fig. 4, female flower furth advanced, with one abortive carpel. Fig. 5, vertical section of same. Fig. 6, the same, still further advanced; *a., a.*, anterior and smaller lobes; *b., b.*, posterior lobes; *c.*, membranous bracts, few and scattered in fig. 1, obsolete in fig. 2, united in a cup in figs. 3, 4, 5 and 6, but cut back in 4 and 5; *d., d.*, hornlike processes growing on the upper part of the carpel and perhaps helping to shelter the emerging embryo; *e.*, style of abortive carpel; *f.*, style of fertile carpel; *g.*, embryo with plumule emerging from fruit; *j.*, horny tissue of carpel, which becomes the comb; *k.*, embryo in earlier stage; *l.*, albumen surrounding lower part of embryo. Fig. 7, the quadripartite comb. Fig. 8, embryo. Fig. 9, embryo further developed; *h.*, cotyledon; *i.*, plumule; *m.*, hypocotyl. Fig. 10, male flower enclosed in leaf-sheath (long-leaved variety). Fig. 11, transverse section of anthers. Fig. 12, anthers opening. Fig. 13, a pollen-cell. Fig. 14, young plant rising from the comb and rooting itself. Fig. 15, branch with female flower at summit. Fig. 16, transverse section, showing anthers dehiscing in three valves and practically one-celled through absorption of the partitions. [Black, J. M. (1913).]

The expression "viviparous" is sometimes loosely used in describing such plants as the Grass, *Poa alpina*, L., in which leafy shoots are found sprouting from the inflorescence. But this employment of the term is inaccurate, since

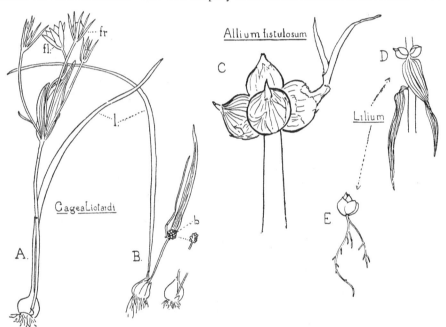

Fig. clvi. Vegetative reproduction. A and B, *Gagea Liotardi*, Schult.; Riffelalp, June; A, flowering plant, outer scales removed from bulb; *fl.*, flower; *fr.*, fruit; *l.*, cylindrical leaves (× ½); B, plant with head of bulbils, *b.*, replacing flowers; in sketch to the right, the outer scales are removed from the bulb (× ⅓); C, *Allium fistulosum*, L., inflorescence replaced by a head of bulbs, one of which is sprouting; Cambridge Botanic Garden, July 20; D and E, *Lilium* (garden hybrid, *tigrinum*, Ker-Gawl. × *Fortunei*, Lindl.) with axillary bulbils; those in the lower leaf-axil have fallen (× ½); E, bulbil such as those shown in D, which has rooted on being kept in damp moss (× ½). [A.A.]

here we have to do with a purely vegetative development of the inflorescence axis, and not with the germination of seeds while still *in situ* in the ovary. It is quite common among Monocotyledons to find examples of "inflorescences" in which the flowers are replaced by tubers or bulbils. *Gagea Liotardi*, Schult., is a well-known instance (figs. clvi, A and B). When examining a number of these plants some years ago at the Riffelalp, I noticed that most of those which bore bulbils were smaller than the flowering plants, so that the bulbil-inflorescences were often found at about the level of the ground, instead of being raised into the air, as are the flowers. In the case of the head of bulbs of the Welsh Onion, *Allium fistulosum*, L., drawn in fig. clvi, c, one of the bulbs has begun to sprout. Instances of this replacement of sexual repro-duction by vegetative budding might be multiplied, but it will perhaps suffice to add that aquatics seem particularly liable to this change[1].

Up to this point we have treated the reproductive shoot as if it were an isolated phenomenon; it now remains to consider the reproductive phase in relation to the general life-history of the plant. This general life-history does not follow a uniform course, but, in the life of the plant, as in that of man, there is a steady crescendo to a period of maximum vitality, followed by an equally inevitable diminuendo. Growth periodicity, so far as it affects the vegetative organs, is more readily observable in Monocotyledons than in Dicotyledons, because, in the latter, this phenomenon is liable to be masked by secondary thickening. Schoute[2] has shown, for instance, that in *Pandanus* the axis forms an elongated spindle, since the diameter of the apical cone, and hence of the stem which it produces, increases to a maximum, and then diminishes. The length of the internodes and the length of the axillary buds show a corresponding periodicity, which, however, lags behind that of the axis-diameter. Observations such as these, which depend on data drawn from a mature plant, present comparatively little difficulty; but records of complete life-histories, from youth to senescence, can only be made by those who are prepared to pursue the matter uninterruptedly over a number of years, and the result is that we possess distressingly few data. There are, however, some notable exceptions. Braun[3], for instance, has carefully followed the development of *Asparagus officinalis*, L. He finds that this plant begins to flower in the third year, but does not reach its full vigour until the fourth or fifth year; from this time onwards it enjoys a period of maturity lasting about fifteen years, after which it suffers decline. I have discussed elsewhere[4] the behaviour of *Elodea canadensis*, Mich., in this country, and the life-history of *Phragmites communis*, Trin., studied by Pallis[5] in the reed swamps of the Danube. The recorded observations upon the life of *Elodea*, *Phragmites*, *Pandanus* and *Asparagus*, seem to suggest the idea that each sexually-produced individual has a definite and limited capacity for vegetative development; its growth rises to a maximum, which it is unable to exceed, and which is followed by a gradual diminution, culminating in old age and death.

[1] Arber, A. (1920[4]), pp. 224—5.   [2] Schoute, J. C. (1907).
[3] Braun, A. (1851).   [4] Arber, A. (1920[4]), pp. 211 *et seq.*
[5] Pallis, M. (1916).

As far as vegetative growth is concerned, the existence of this definite periodicity can only be ascertained by detailed observation. But that there is often a certain periodicity in flowering is much more obvious. It is a well-known fact that Bamboo forests over large areas in the East, after being purely vegetative for many years, may break simultaneously into flower in a single season, and then die. The effects of this sudden flowering may be of a startling nature. It is recorded, for instance, that a species of *Melocanna*, which occupied 6000 square miles in a certain district in India, flowered and fruited to such an extent in 1867, that a survey officer, who was dealing with the region, was obliged to cease work some weeks before his task was concluded, owing to the breakage caused among his theodolites and plane tables by the fall of the heavy fruits. After flowering, the plants all died, and their place was taken by countless seedlings, which eventually formed a uniform and almost pure jungle of *Melocanna*[1].

It has been customary to attribute gregarious flowering, of the type just described, to external influences (e.g. drought), but the view seems to be gaining ground that it is the result, rather, of inherent factors; those holding this opinion suppose that this long-deferred flowering is the inevitable manifestation of the attainment of sexual maturity, and that it occurs simultaneously in countless individuals, because they all date from the last seeding period, possibly many years earlier[2]. This view is confirmed by the fact that, in the case of a species of *Phyllostachys*, about which records exist, no efforts have hitherto been successful in preventing flowering, when the due time has arrived[3]. This habit of certain Bamboos—simultaneous flowering, followed by death of all the individuals over a large area—is a matter of painful interest in the countries where they abound, because these plants are used, not only in every kind of construction, from building downwards, but also for food. Po Chü-i, a Chinese poet, wrote, in the first half of the ninth century[4]:

> "My new Province is a land of bamboo-groves:
> Their shoots in spring fill the valleys and hills.
> The mountain woodman cuts an armful of them
> And brings them down to sell at the early market.
> Things are cheap in proportion as they are common;
> For two farthings, I buy a whole bundle.
> I put the shoots in a great earthen pot
> And heat them up along with boiling rice.
> The purple nodules broken,—like an old brocade;
> The white skin opened,—like new pearls."

So important, indeed, are the Bamboos in the life of man, that it is not surprising that records regarding the flowering of *Phyllostachys puberula*,

---

[1] Stapf, O. (1904).   [2] Seifriz, W. (1923).   [3] Kawamura, S. (1911).
[4] *A Hundred and Seventy Chinese Poems*. Translated by Arthur Waley. London, 1918.

Munro, occur not only in modern writings, but also in ancient manuscripts of China and Japan. A Japanese botanist[1], who has brought these data together, shows that flowering is known to have taken place in this species in the years 292, 813, 931, 1114, 1247, 1666, 1786, 1848 and 1908; the intervals between these flowering periods are mostly about sixty, or multiples of sixty years, thus conforming to a definite numerical cycle. In *P. bambusoides*, Sieb. et Zucc., on the other hand, the life-cycle is said to be 120 years. As another example we may take the climbing Bamboo of Jamaica, *Chusquea abietifolia*, Griseb.[2], which flowered in 1885—6, and then died down everywhere, and was succeeded by millions of seedlings; it is remarkable that plants which had been transferred to Kew Gardens flowered in the same year as the wild Bamboos in Jamaica. In *Chusquea abietifolia* the next flowering period was 1918, so the life cycle seems to be about 32 years.

A rhythmic alternation of vegetative and sexual phases is not often so conspicuous as it is in the life-history of the Bamboos, but in many Monocotyledons the development of shoots and roots as storage organs leads to another peculiar form of periodicity, in which a rhythm is established between periods of normal growth and periods of tuberisation. We have already considered this subject in connexion with the Ophrydeae, in which root-tuberisation due to fungal infection alternates with a non-infected differentiating period, in the course of which the foliage-leaves, the inflorescence, and the bud for next year are formed (pp. 29, 30). It is probable that tuberisation is also dependent on fungal infection in many other cases, in which this influence has not yet been proved. But, whether due to this cause or not, a telescoping of the main axis, giving rise to the geophilous habit, in which the underground organs assume special importance, is highly characteristic of Monocotyledons. The storage of carbohydrates and other food material, which is a common feature of such geophytes, leads to certain peculiarities in the succession of phases in the life-history, which we must now consider.

One of the outstanding results of the accumulation of reserves in an abbreviated shoot is a capacity for the production of flowers at a time when no assimilating organs are in active work; *Nerine curvifolia*, Herb., shown in fig. clvii, p. 208, will serve as an example. The umbel of brick-red flowers, mingled with delicate, almost hair-like bracts, is borne on an axis which produces no foliage-leaves. But on dissecting the bulb, which is formed of succulent leaf-bases, we find, beside the base of the inflorescence axis, the group of close-packed, distichous leaves, which will develop later (fig. clvii, B).

A separation between the flower and leaf phases, similar to that in the bulbous *Nerine* just described, is to be observed in the tuberous *Sauromatum guttatum*, Schott, an Indian Aroid, sometimes sold in this country under

---

[1] Kawamura, S. (1911).   [2] Morris, D. (1886), and Seifriz, W. (1920).

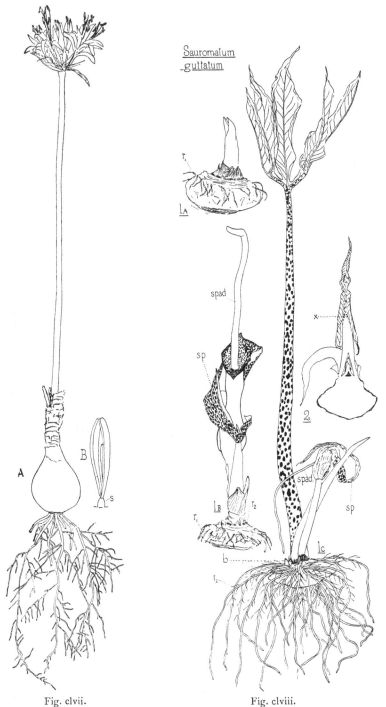

Fig. clvii.   Fig. clviii.

Fig. clvii. *Nerine curvifolia*, Herb., var. *major*. A, plant (× ¼), September 9; B, leaves for next season dissected out of the bulb; *s.*, leaf-sheaths which will form future bulb (× ¼). [A.A.]

Fig. clviii. *Sauromatum guttatum*, Schott, Monarch-of-the-east. Fig. 1 A, tuber, May 5, 1923, with roots of 1922 (r_1) in shrivelled condition (× ¼); fig. 1 B, similar plant flowering, June 1, 1923 (× ¼); tuber shrunken; r_2, rudiments of this year's crop of roots; *sp.*, spathe; *spad.*, spadix (× ¼); fig. 1 C, plant at leafy stage, June 25, 1923 (× ¼). Fig. 2, non-flowering plant, June 25, 1923 (× ¼); from level × downwards, the plant is cut in two longitudinally, to show apical bud enclosed in base of pseudo-terminal leaf. [A.A.]

the name of Monarch-of-the-east. The corm is shown in fig. clviii, 1 A, as it appears early in May; last year's roots, r_1, are dead and shrivelled. If it is kept in a sunny place, but without earth or water, in a few weeks it puts up the large bizarre inflorescence with its spotted spathe drawn in fig. clviii, 1 B; if it is then planted in earth, the decorative leaf shown in fig. 1 C is promptly produced, and a large crop of roots appears. These roots are strongly contractile, and it has been recorded that, if the tuber is planted at the surface of the ground, a search two months later will reveal the fact that it has descended to a depth of six inches, while the roots have shortened to about half their original length[1].

One of the features which seems to be correlated with the geophilous habit, is an extreme slowness in reaching maturity. That this should be so is not surprising, when we remember that tuberisation is, in a sense, pathological; it represents a slowing down of development, associated with the storage of that food material which would normally have been used up in the process of differentiating growth. Our native Wild Arum, *Arum maculatum*, L., is an instance of this tardiness in coming to maturity[2]. Seeds which I sowed one September were dug up in the following April, when the cotyledon-tip was found to be still embedded in the seed, while the hypocotyl had developed into a small tuber (*t.*, fig. cxlv, 4 C, p. 190). In the following Autumn, the stage shown in fig. 4 D was reached, and such a condition as fig. 5 B is not attained until the following Spring—that is to say, the young plants, from seeds which ripen in September, do not make their appearance above the soil until considerably more than a year from the time of sowing. The roots are contractile, ultimately pulling the tuber down to a depth of about 10 cms.; fig. cxlv, 2 B, shows the length, in proportion to the spathe, of the inflorescence axis, the greater part of which is subterranean. It is said that *Arum maculatum* rarely flowers before its seventh year. Unfortunately it is only in a few cases among geophytes that we have records of the age at which flowering first occurs; the bulbous Irises, for instance, seldom, if ever, flower before their fourth or fifth year[3], while, in the Fawn-lily, *Erythronium americanum*, Ker-Gawl., a five-year cycle from seed to flowering plant has been described[4]. The Florist's-tulip furnishes perhaps the extremest example of a leisurely life-history. Seedling Tulips take some years to reach the flowering stage, and they are then capable, in certain instances, of remaining for more than fifty years as self-coloured "breeders," after which they "break"—that is to say, take on the more varied colouring characteristic of the adult form[5]; the breeder may be regarded as a juvenile form prolonged out of all conscience.

[1] Scott, R. (Mrs D. H.) (1909).
[2] Rimbach, A. (1897²); Scott, R. (Mrs D. H.) and Sargant, E. (1898).
[3] Dykes, W. R. (1913).   [4] Blodgett, F. H. (1900).   [5] Solms-Laubach, H., Graf zu (1899).

CHAPTER IX

TAXONOMY AND ITS INTERPRETATION

THE existing classification of the Angiosperms into species, genera, and families, and the grouping of these families into larger units—cohorts or orders—has been the fruit of long years of unremitting labour on the part of a legion of botanists. The results attained, though, as time passes, they are inevitably modified in detail, have yet shown a stability which is a striking tribute to the work done by systematists as a body, and also to the special genius for the perception of affinities possessed by the more outstanding of the men who have devoted themselves to this work. It is remarkable that—although, until recently, the criteria used in Systematic Botany have been based almost entirely on external morphology—studies on the anatomical and cytological characters of the various groups, made with the aid of modern laboratory technique, have, in general, merely confirmed the conclusions of the taxonomists. Since this is the case, we may, in the present chapter, accept the system as representing a sufficiently close approximation to the truth to serve as a basis for analytical study.

During this study of Monocotyledons, the point that has chiefly struck me about generic and specific differences, is the impossibility of explaining them as adaptive. This is of course no new idea; in recent years it has been forcibly expressed by Willis, Gates and others[1]. The fact that it is, in many cases, practicable to identify genera and even species from the anatomy of leaves or peduncles alone, and that the differences used as criteria cannot, by any stretch of imagination, be supposed to be of special utility to the plant, seems to put any adaptational explanation completely out of court[2]. Nor is there any evidence that the conditions of life have a direct causal relation to these or other specific differences, and there thus seems no escape from the view that they are the result of spontaneous mutations due to inherent factors. If this be so, the same interpretation must also be accepted, not only for varietal differences, but also for generic differences, and even for the characters differentiating larger groups; for it is a remarkable fact that in all these cases, the individual differentiating features are of the same type. As a concrete example, we may recall that one of the characters distinguishing the different species of the Liliaceous genus *Trillium*, the Trinity-flower, is the presence or absence of a petiole. *T. grandiflorum*, Salisb., and *T. sessile*, L., have a whorl of sessile leaves below the flower,

[1] Willis, J. C. (1923); Gates, R. R. (1918), etc.
[2] Gatin, V. C. (1920); for a consideration of water plants from this point of view, see Arber, A. (1920[4]).

while in *T. recurvatum*, L., the leaves are stalked (cf. figs. clix, 2, 4, 3); but varieties of Wake-robin (*T. grandiflorum*) sometimes occur in which the leaves are petiolate. This group, moreover, provides illustrations of the identity between varietal and generic differences. One of the distinctions between *Paris* and *Trillium*, is 'that *Paris* has a 4-merous flower, while *Trillium* is 3-merous. But *Paris quadrifolia*, L., is sometimes found with six foliage-leaves below a 3-merous flower (fig. clix, 1 B), while the Trilliums may have their leaves and the parts of their flowers in fours[1]. A more unexpected feature, however, is that differences found within one species

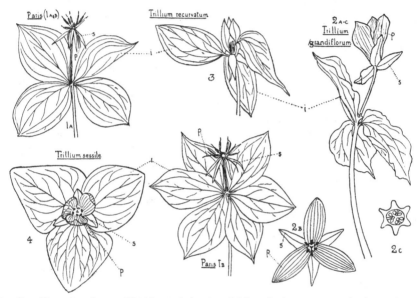

Fig. clix. Flowering shoots of Parideae; *i.*, involucral foliage leaf; *s.*, outer perianth member; *p.*, inner perianth member. Figs. 1 A and B, *Paris quadrifolia*, L.; fig. 1 A, normal flowering shoot (× ⅓); fig. 1 B, six-leaved shoot (× ¼), Caterham, June. Figs. 2—4, *Trillium*; figs. 2 A—C, *T. grandiflorum*, Salisb., flowering shoot (× ¼); fig. 2 B, flower from above; fig. 2 C, transverse section of ovary. Fig. 3, *T. recurvatum*, Beck, flowering shoot (× ¼). Fig. 4, *T. sessile*, L., flower and involucre seen from above (× ⅓). [A.A.]

may be of the same character as differences which are used to distinguish the largest groups. Fyson[2], for instance, in his study of *Eriocaulon*, has pointed out that, in certain species the number of sepals may be either two or three, and he has emphasised the significance of this fact; it is indeed surprising that a numerical difference, which is one of the diagnostic characters for the two principal divisions of Angiosperms, is here found within the limits of a single species. Specific characters, again, may be identical with those used in diagnosing larger classificatory units; the fruits of different species of *Yucca* may be septicidal capsules, loculicidal capsules,

[1] Gates, R. R. (1917).   [2] Fyson, P. F. (1921).

or berries, and yet these distinctions in the type of fruit have been used in the Liliaceae—to which *Yucca* itself belongs—to distinguish sub-families[1]. The same story is repeated when we turn to differences between genera. Rendle[2] has pointed out that both hypogyny and epigyny are found within the limits of the very natural family Bromeliaceae, although, among Dicotyledons, this character is found useful in separating large groups. The difference between endospermic and non-endospermic seeds, again, is used as a diagnostic character of large groups, and yet in related genera among the Aroids both types occur[3]; in the Gramineae, also, although the seeds are so characteristically endospermic, the genus *Melocanna*[4] has fruits in which the endosperm is reduced to a structureless film between the pericarp and scutellum[5].

I think that the upshot of these facts is that the consideration of individual characters—regarded as detached entities—contributes little towards an understanding of the evolution of the Angiosperms. We must approach the subject on a different line, and proceed from the family to the species, rather than from the species to the family. With this in view, it may be well to concentrate on some one group of plants, which seems to be a homogeneous and natural one, and to try the effect of passing from the organs to the characters, instead of proceeding in the opposite direction, as we do when the characters form the basis of the discussion. For this purpose I propose to consider the Araceae, because they combine great naturalness as a group with an amazing range of variation, and because Engler's work[6] on the family not only supplies the information we want, but also subjects the data to a luminous analysis.

The Araceae are so universally recognised as forming a really natural group, that it is perhaps scarcely necessary to labour the point. I will only mention that Dalitzsch[7] has demonstrated that spongy parenchyma of a special and peculiar form, in which the cells are arranged to make a series of fluted columns, is characteristic of most Aroid leaves. Since this is a minor structural feature, lying outside the usual range of systematic criteria, it furnishes a strong confirmation of the naturalness of the family. If we accept the essential homogeneity of the group, the next step is to analyse such heterogeneity as is found within its limits. This has been comprehensively done by Engler, who has traced progressive lines of differentiation, for the inflorescence as a whole, and for its constituent parts, and also for a number of vegetative features. I cannot, for lack of space, cite his results in full—

---

[1] Baker, J. G. (1875).   [2] Rendle, A. B. (1904).
[3] Engler, A. (1881—1884).   [4] Stapf, O. (1904).
[5] Willis, J. C. (1923) gives a number of examples of the same type; the above discussion was, however, written without reference to Willis's paper.
[6] Engler, A. (1881—1884).   [7] Dalitzsch, M. (1886).

as I should wish to do; in the following paragraphs, certain of them are summarised with extreme brevity.

Branching.

The branching, in a few instances, is monopodial, with budding from every leaf axil; but it is sympodial in the majority of the Araceae, in which the system becomes limited by certain lateral buds being, as it were, favoured; in some cases these favoured shoots produce an indefinite number of leaves, while in others their number is strictly limited.

General relations of spathe and spadix.

There is a range from a simple open form of spathe, resembling a foliage leaf, in which there is complete independence of spathe and spadix, to more complex forms, in which the spathe is less leaf-like, and its relation with the spadix is closer, and finally to forms in which there are actual fusions, of more than one type, between spadix and spathe.

Flowers.

A series can be traced according to whether the spadix bears:

(i)   hermaphrodite flowers,
(ii)   male and female flowers,
(iii) male, female, and sterile flowers,
(iv)   male, female, sterile, and rudimentary flowers[1].

The plants bearing unisexual flowers are generally monoecious, but dioecism is attained in certain rare cases.

There is also a reduction series as regards the number of flowers, particularly in the female part of the inflorescence. The number of female flowers may be reduced to a few, and ultimately to one. We find the last terms of this series in the related Lemnaceae, which are doubtless the most reduced representatives of the Araceae. In *Lemna* and *Spirodela* the spathe includes one monocarpellary female flower, and two male flowers, each of which is reduced to a single stamen; in *Wolfiella* reduction has gone so far that the spathe is absent, and the inflorescence consists of a single stamen, representing one male flower, and one gynaeceum, representing a female flower.

Perianth.

In the perianth there may be a progression from free members in two whorls to united members fused into a single organ. Or the perianth may be lost altogether.

---

[1] E.g., *Arum maculatum*, L., see p. 189 and fig. cxlv, p. 190.

Androecium.

The stamens may show reduction from two whorls to one, and also reduction from the number 6, to 5, 4, 3, 2, 1. Another progression, from free filaments to filaments united in some degree, culminates in a synandrium[1]. Corresponding progressions can be traced for the staminodes.

Gynaeceum.

The gynaeceum according to Engler has two or more equally developed carpels, which may form either an equivalent number of loculi, or else one loculus; or two or more carpels, of which only one is fully formed; or one median carpel[2]. In all these cases there is a range from many ovules to one.

Seeds.

The seeds are generally endospermic, but endosperm is sometimes absent.

It is clear from this summary of the system of variation, that, in the Araceae, a few distinct tendencies run through the modifications of the different organs. The branching, for instance, shows a gradation from an indefinite system, with no limitation in the number of branches or leaves, to a limited system, in which only certain buds develop, and in which the number of leaves produced on the shoots is few and rigidly standardised. The branching may thus be said to constitute a *reduction* series. In the relation of spathe and spadix, on the other hand, the factor which chiefly affects the result is the tendency to *fusion* between these organs. This may seem, at first glance, to be a somewhat different thing, but it must be remembered that fusion is not really the active process that its name might seem to imply, but would generally be more accurately described by the phrase "failure to separate." In the flowers themselves, we find *reduction* in number, and also reduction from hermaphroditism to unisexuality. Another factor, which also supervenes, is *degeneration*, which gives rise not only to sterile flowers, but even to undifferentiated rudimentary flowers, which never become separate, and which form a continuous coating to the spadix. The progression in the perianth is characterised in one direction by fusion, and in another by loss, while in the androecium, again, both reduction and fusion are the operating factors. In the gynaeceum, reduction, in the number of both carpels and ovules, is principally responsible for the lines of progression.

When we look broadly at these results, we seem to be justified in deducing from them that the archetype of the Araceae was a plant with unlimited monopodial branching, whose inflorescence axis bore a bracteole (spathe), and was clothed, above the spathe, with bractless, sessile, hermaphrodite

---

[1] E.g., *Spathicarpa sagittifolia*, Schott, fig. cxxxviii, 6, p. 181, and p. 191.

[2] These conceptions may have to be modified; see foot-note 1, p. 215.

flowers, pentacyclic[1] and trimerous, with free inferior perianth members and stamens, and numerous endospermic seeds. From this type, the immense variety met with amongst the present-day Aroids has been derived, chiefly by the operation of three factors, *reduction, fusion* and *degeneration*. It is remarkable that these are the very factors which—long before the general recognition of evolution—A. P. de Candolle[2] named as those which, in each family, "tendent à altérer la symétrie primitive." He also included in his enumeration of these factors, *multiplication of parts*, which plays an important rôle in some families, but which does not seem to achieve expression in the particular case of the Araceae.

But though reduction, and the related factors, fusion and degeneration, have undoubtedly been the principal elements in the differentiation of the Araceae, other factors, of a somewhat different kind, must not be overlooked. Engler points out that certain minor histological features serve as criteria in classification, e.g. the presence or absence of tubular tannin cells, spicular cells, and latex vessels. That these features distinguish different cycles of affinity, and are not related to a common mode of life, is indicated by the fact that climbers belonging to the genus *Pothos* have no latex tubes and no spicular cells; climbers belonging to *Monstera* and related genera have spicular cells; while climbers belonging to *Philodendron* have latex tubes, but no spicular cells. It is clear that we are here concerned with special by-products due to some variation in the plant's metabolism, and we must hence add the factor of *chemical change* to those already recognised as playing a part in differentiation. Unfortunately we have at present no full knowledge of the relation of chemistry to morphology, but it is probable that many specific differences in anatomy, such as those which I have illustrated for the *Juno* Irises (fig. lxvii, p. 90), in which the relative development of fibres, the characters of the cell-wall, etc., play a part, will eventually be found to have a chemical basis. The possibility is by no means excluded that many such chemical changes may ultimately prove to be reduction phenomena, like so many of the morphological changes which we have been studying. This is suggested by the work of a bio-chemist, who has recently claimed that the absence of starch in the leaves of many Monocotyledons is due to the lack of one or more of the necessary carbohydrate enzymes, while in the Dicotyledons the full enzyme apparatus is produced[3].

When we turn to the leaf of the Araceae, we find ourselves in the presence of a progression belonging to a somewhat different order, but which, nevertheless, probably owes its first impulse to the effects of reduction. There are a few instances of simple leaves with an isobilateral equitant limb, but the great majority of Aroids show a differentiation into "blade" and petiole.

---

[1] If the view of the gynaeceum suggested in Saunders, E. R. (1923) proves generally applicable to Monocotyledons, we must assume the primaeval Aroid to have been hexacyclic; see pp. 197—199.

[2] Candolle, A. P. de (1827). [3] Chapman, R. E. (1924).

The form of the blade may be simple or subdivided, and extreme complexity may be reached; perforations[1], due to local necrosis, and other minor peculiarities also occur. If the phyllode theory, discussed in an earlier chapter (p. 100), be accepted, the equitant leaf will be interpreted as a simple, bladeless, petiolar phyllode, while the more elaborate forms represent the evolution of different types of pseudo-lamina from the distal region of the phyllode. Engler notes that, in the Araceae, the nervation of leaves of similar outline, but belonging to plants of different cycles of affinity may conform to distinct types. I think this is rather what one would expect on the phyllode theory, since the pseudo-lamina was probably evolved independently on various lines. I will not pursue this subject farther, except to suggest that we seem here to be in presence of another factor in family-differentiation—*the elaboration of a replacement organ*, formed from an existing structure, but taking the place of one which has been lost. An analogous case may be found in the petaloid development of the stamens, which assumes the place of the absent perianth in *Potamogeton*.

A farther factor, which, though it can hardly be described as a *cause* of specific differentiation, yet certainly facilitates it, is *seasonal dimorphism*; this phenomenon can also be illustrated from among the Araceae. *Arum italicum*, Mill., which occurs as a rarity in Britain, closely resembles *A. maculatum*, L., but differs from it in the seasonal rhythm of its life-history; it produces its leaves in the late Autumn and Winter, while those of *A. maculatum* do not appear until the Spring. It has been suggested by Buchenau[2] that *Spiranthes aestivalis*, Rich., and *S. autumnalis*, Rich., are another pair of species which may come into the same category, and he also regards *Triglochin laxiflora*, Guss., as a "petite espèce," separated from *T. bulbosa*, L., by alteration in the flowering time. This seasonal dimorphism might well lead to the establishment of new species, since two forms, of which the flowering time has ceased to synchronise, are as effectually isolated as by a geographical barrier.

Though the Aroids form a specially favourable group in which to trace the course of differentiation from the family type to the specific type, any other large family, or cohort of closely allied families, might be studied on the same lines. In all the great families we find certain plants which are more generalised than the rest in floral or in vegetative construction, and which appear to give indications of the nature of the primaeval family stock; and lines of progression may be discovered, indicating the general routes, from uniformity to diversity, by which the specific forms of the present day have travelled[3]. On comparing the various cohorts from this point of view,

---

[1] E.g., *Monstera deliciosa*, Liebm., p. 72 and figs. xlvi, 2 A—C, p. 69.   [2] Buchenau, F. (1896).
[3] Some of the variations met with in the Liliiflorae have been indicated in pp. 184—188 and figs cxliii, p. 186, and cliv, p. 202.

one cannot but be struck by the fact that, despite the amazing variability displayed, the limits of the group-type are never overstepped. The Aroid, for instance, though suffering unending modifications, yet remains recognisably an Aroid to the last[1]. Considerations of this kind have led me to think that the great groups of Monocotyledons have not achieved unlikeness by divergent modification, but that they must have been of different types from the moment of their appearance. They seem to resemble a series of *motifs*, each of which, however much modified, and however variously orchestrated, remains recognisable and individual. We have no evidence from Palaeobotany for the former existence of synthetic types uniting any of the Monocotyledonous cohorts, and I am inclined to suppose that these great groups will ultimately be traced back to a very remote antiquity, without displaying a common origin. I thus feel that the speculations regarding the relations of the cohorts, which have been suggested by various botanists[2], can, in the nature of things, have little value. One may sometimes be able to hazard a reasonable guess as to the origin of a species or a genus, or even of a minor family, but, as soon as we go beyond this, we lose our grip on reality. It seems to me that the reluctance to express any decided opinion upon the relation between the cohorts, which systematists—unlike laboratory botanists—have almost invariably displayed, is fully justified. It is true that there are families that seem, at first glance, to furnish links between the larger groups, but their claim seldom outlasts a critical examination. This is illustrated by the case of the Triuridaceae—a group of extremely reduced saprophytes, now classed with the Monocotyledons, but which have sometimes been thought to form a link between the two main classes of Angiosperms. Ethel Sargant, in an unpublished note, reviewed our data regarding the family, and concluded, "The doubt as to its systematic position is due then to our ignorance of its characters or to their obliteration, and no more presupposes a structure intermediate between that of Monocotyledons and Dicotyledons than an inscription so defaced that it might equally well be Spanish or Italian could be considered as a link between those languages."

That the original synthetic stocks of all the great Monocotyledonous groups will ultimately be shown to have been derived from a single primaeval type of extreme antiquity—the Ur-monocotyledon—is perhaps the most probable supposition, but it is far from being proven; indeed there seems at present no hope of penetrating the mists that shroud the origin of the cohorts. This distinction between the origin of the groups—which lies

---

[1] Rendle, A. B. (1904) has drawn attention to the Orchidaceae as an example of the great diversity which is possible in flowers constructed on one uniform plan.

[2] E.g., Delpino, F. (1896); Fritsch, K. (1905); Lotsy, J. P. (1911); the last title is cited because this book contains an epitome of the views of Hallier and others, but it should be noted that in a more recent work (Lotsy, J. P. (1916)) the author has concluded that the effort to trace phylogenies is a vain one.

beyond our ken—and their subsequent differentiation—which is amenable to study—is not a new one. It has been applied to plants by Guppy[1], while, on the animal side, a corresponding generalisation has been made by Depéret, who postulates "two distinct *mechanisms*" in the evolution of extinct beings, one of which is intermittent, and gives rise to new branches, while the other, which is "continuous and, so to speak, normal," is responsible for the slow subsequent development of these phyletic branches[2]. Our analysis of Engler's study of the Aroids (pp. 212—216) confirms the de Candollean view that it is reduction, and related factors, which provide the key to the origin of the specific forms which now represent the various families. I wish to suggest that this is not merely fortuitous, but is due to the fact that the evolution of these families is now in a down-grade period. It is not unreasonable to suppose that groups, like individuals, enjoy a juvenile, up-grade phase, followed by a mature period, which passes insensibly into senescence, and in which, owing to the gradual diminuendo of vitality, structural reduction plays the chief part. It seems to me that it is the differentiating features of down-grade evolution with which Mendelian analysis is exclusively concerned, and that this is the reason why it is useless to expect direct light on constructive, up-grade evolution from the work of the geneticists.

The meaning of reduction-processes in evolution has been rendered unnecessarily obscure by the misleading terminology coined by morphologists. The use of the word "economy" in connexion with reduced features in flower construction, has been peculiarly unfortunate, because it has carried with it the unwarranted implication of purposefulness. No one would claim that the loss of hair in the elderly human being is an adaptation for the purpose of economising in brushes—on the contrary, we recognise it at once as a symptom of failure in vitality due to age. And it seems reasonable that the same common-sense explanation should apply to racial losses and reductions. I take it that the epoch in which the cohorts of Monocotyledons took their origin, represents the active, constructive, juvenile period of Angiospermic evolution, and that we are living to-day in the period of their late middle life, in which no energy is left for changes other than those which are essentially losses or limitations.

That the origin of the principal subdivisions of Monocotyledons should be sought for farther and farther back in time, and that the progress of research should emphasise their independence, rather than any connexion between them, is in harmony with the modern view of botanists as to the age of all the great groups. D. H. Scott[3] now recognises the Seed-plant phylum as an independent line, which has run its own course from Devonian times onwards, and he considers that the earlier view, which derived the

[1] Guppy, H. B. (1919).   [2] Depéret, C. (1909).
[3] Scott, D. H. (1920).

Pteridosperms from the Ferns, can no longer be maintained. F. T. Brooks[1], again, in dealing with quite another form of vegetation, holds that the Fungi are not derived from the Algae, but "are a monophyletic group, established in far-distant times from protist organisms, which have evolved along their own lines just like any other group of plants or animals." In considerations such as these, analogies drawn from the animal kingdom are of special value; it is, for instance, significant, that the view that the lower, non-placental Mammals and the higher Mammals were derived from a common ancestor, has ceased to hold the field, and it is recognised that to postulate a common origin for these two groups is a matter of pure hypothesis[2].

It thus seems that we must, at least for the moment, give up the hope of bridging the gulfs which separate the great Angiospermic groups. But even if this be admitted, it is still open to us to attempt an analysis of the features on which their distinctness depends. In embarking on this, however, we are met at once by the difficulty that no definition by detached characters is of much value, when we are trying to understand how the difference between the groups is determined. For we have shown (pp. 210—212) that the importance of characters, considered individually, varies from group to group, so that features used to separate two great groups, may elsewhere merely serve to distinguish two forms of a single species. But, if detached characters prove of little use to us, in what direction are we to look for help in visualising the essential differences between the main classificatory units? It seems to me that it is in the writings of Bergson, rather than in those of professed biologists, that we find the clue to the problem. Bergson was not, of course, concerned with our particular question, but, speaking of evolution in general, he says: "There is no manifestation of life which does not contain, in a rudimentary state—either latent or potential,—the essential characters of most other manifestations. The difference is in the proportions....In a word, *the group must not be defined by the possession of certain characters, but by its tendency to emphasise them*[3]." The value of this dictum will be realised by applying it to some concrete case. We may take, as an example, the trimerous character of the flower, which is prominent as a systematic mark of Monocotyledons. This trimery is, however, by no means universal, and the change from one numerical system to another may occur within one small group[4]. In the genus *Aspidistra*, for instance, the species *A. typica*, Baillon, is trimerous with nine perianth members, while the others are tetramerous with eight[5]. Again, in *Paris quadrifolia*, L., which normally has four perianth members and eight stamens (fig. clix, I A, p. 211), examples may be found with six perianth members and twelve stamens (fig. clix, I B); conversely, various cases of trimery in Dicotyledons might be cited. It is thus clear that the

[1] Brooks, F. T. (1923).   [2] Depéret, C. (1909).   [3] Bergson, H. (1911).
[4] See p. 211.   [5] Baillon, H. (1894²).

character of trimery cannot be claimed as furnishing an absolute distinction between Monocotyledons and Dicotyledons. But, expressing the matter in terms of Bergson's generalisation, we should say that the tendency to trimery exists in both classes, but while it *is subordinated* in Dicotyledons, it *preponderates* in Monocotyledons.

But though the conception of tendencies rather than states is helpful in morphological considerations, its full value is not revealed until we look at our data from the chemical standpoint. One of the most important distinctions between the metabolism of Dicotyledons and that of Monocotyledons, is that whereas in Dicotyledons a temporary reserve of starch is freely formed in the leaves during photosynthesis, in many Monocotyledons little or none is produced. We owe to F. F. Blackman[1] a suggestive study of this subject, based upon the data of Meyer[2], who, forty years ago, examined a large number of Angiosperms from the point of view of the amount of leaf-starch present. Meyer's work is by no means exhaustive, and it is much to be desired that some botanist, with access to a botanical garden, would treat the subject with greater fulness, and with the use of critical comparative methods—but in the meantime we may accept the existing data as trustworthy, at least in outline. Meyer classifies the Angiosperms into six sections, ranging from Division I, in which starch is richly present in the leaves, to Division VI, in which there is none at all. His results do not reveal a hard-and-fast distinction between Dicotyledons and Monocotyledons, but they show that, whereas most Dicotyledonous families fall into Divisions II and III, the majority of Monocotyledons occupy positions in the later divisions, and none of them attain to Division I. Blackman attributes this difference to the fact that, in Monocotyledons, a high critical sugar concentration is needed to initiate starch formation, but another writer has recently suggested a somewhat different explanation[3]. The details of the interpretation do not, however, concern us at the moment, since they do not affect the main contention—that we are in presence of a certain fundamental chemical character, whose centre of gravity, if we may so express it, shows a shift in Monocotyledons as compared with Dicotyledons. Another broad difference of a similar type between the metabolism of the two classes is indicated by the fact that Monocotyledons are relatively poor in alkaloids[4].

And it is not only between Dicotyledons and Monocotyledons as a whole that certain chemical distinctions may be drawn. When we compare the cohorts Liliiflorae and Farinosae—which in many ways closely resemble one another—we find that the chemical constitution of the endosperm is the principal differentiating factor[5]. In the "horny" endosperm of the Liliiflorae, reserves are stored in the form of thick cell-walls, plasma, and oil (e.g. *Tamus*

[1] Blackman, F. F. (1921).   [2] Meyer, A. (1885).   [3] See p. 215.
[4] Onslow, M. Wheldale (1911).   [5] Engler, A. (1892).

communis, L., frontispiece), while that of the Farinosae is "floury," the cell-walls being thin, and the reserves taking the form of starch-grains. But this, like the other chemical distinctions to which we have referred, is not absolute, for the Philydraceae (Farinosae) are described as having "albumen subcarnosum."

There is evidence, again, that some chemical peculiarities may be repeated in more than one genus in a family: saffron, or some substance closely akin to it, is found, for instance, among the Iridaceae in *Tritonia aurea*, Poppe, as well as in the genus *Crocus*[1]. Certain compounds, on the other hand, appear to be confined to individual genera; veratric acid, for example, seems to be limited to *Veratrum*, while the cyanogenetic glucoside, dhurrin, is peculiar to *Sorghum*[2].

None of those chemical differentiations to which we have referred produce any obvious effect upon the morphology, but there are certain other cases in which a relation between chemistry and form is clearly traceable. Three American workers[3], who have studied succulent plants, have come to the conclusion that their succulence is closely related to pentose formation; "whenever the water content of a cell becomes low, some of the hexose-polysaccharides, which have a low imbibition capacity, are converted into pentosans, which have a high hydration capacity, the action having the force of a regulatory adjustment, and as the change is irreversible, the pentosans are accompanied by a permanent succulence." If this view be accepted, we are, for the first time, furnished with a direct chemical explanation of the chief peculiarities of form met with in a number of Angiosperms. Another case in which chemistry is again an essential agent in form-determination, is that of tuberisation. This subject is discussed in an earlier chapter, in which we have shown how the structure of the Ophrydeae is modified by the influence of the fungus which infects the roots (pp. 26—30). We did not then consider the means by which this influence is exerted, but the work begun by Bernard, and carried to its logical conclusion by a later worker[4], indicates that the influence of the fungus may be replaced, artificially, by that of certain sugars; that is to say, the influence which determines tuberisation, and which thus profoundly affects the form and life-history of the Orchid, is a chemical one. The kind and degree of change which this chemical action can determine, recalls the effects induced by the products of the ductless glands in man. In pathological cases, the excessive or minimal character of these secretions may alter the bodily structure to an extent which appears almost miraculous; and it has even been suggested that the varying personal attributes of the different races of mankind are an ex-

[1] Heim, F. (1896).   [2] Onslow, M. Wheldale (1911).
[3] MacDougal, D. T., Richards, H. M., and Spoehr, H. A. (1919).
[4] Knudson, L. (1922) and (1924).

pression of the varying activities of the glands in question[1]. If this analogy be indeed well-founded, it establishes a close connexion between the phenomena of pathology and of racial differentiation—a connexion which probably finds its counterpart in plants. If tuberisation due to fungal infection occurred rarely and sporadically among the Ophrydeae, we should undoubtedly call it a disease, but, since it is universal, we call it a tribal character. It seems to me that one of the most hopeful fields for the future student of race-history will be the study of pathology in its relation to evolution. And it is to the bio-chemist that he must chiefly look for help in the problems that he will meet in this field, since—in the words of Muriel Onslow—"the plant form is an expression of its chemical constitution[2]."

[1] Keith, A. (1919).   [2] Onslow, M. Wheldale (1911).

CHAPTER X

PARALLELISM IN EVOLUTION

THE phyletic history of the plant world seems to have been visualised by Darwin in the form of a tree, in which the ultimate branchlets stood for the living members, while the trunk and major boughs represented extinct synthetic stocks. But, in the period of more than sixty-five years that has elapsed since *The Origin of Species* appeared, this conception has suffered a change. We have come to realise that there is singularly little evidence for the coalescence of life-lines as they are traced back in geologic time, and there seems a tendency to replace the mental image of the tree by the very unsatisfying one of a bundle of osiers. And it is not only the thinkers of the present day who have rejected the tree metaphor; so long ago as 1884, Nägeli[1] suggested, as a possibility that must be taken into account, that Monocotyledons and Dicotyledons might have originated independently from extremely low types of Vascular Cryptogams, now extinct. When we come to ask ourselves to what factor this change of ideas is principally due, the answer must, I think, be, that we owe it to our growing realisation of the predominant part that parallelism has played in evolution. Although Darwin himself drew attention to the existence of "analogous" variation, the general trend of his work was based on the assumption that resemblance was unimpeachable evidence of common descent. And, so long as the doctrine of Natural Selection held the field, such an assumption was the only logical one. If evolution depended on the accumulation of variations whose direction was in no wise determined, there was small likelihood of the origin of similar forms, except through common descent. This idea penetrated from science to the humanities, and seems to have survived there unimpaired, although, in biology, the dogma has lost much of its cogency. The columns of almost any literary paper furnish examples of the doctrine that the occurrence of similar passages, in authors of successive periods, is evidence for direct filiation of ideas; little heed seems to be paid to the degree to which the very nature of man's mind directs his thought along roads which his predecessors have travelled in the past, and which his posterity will tread again and again. Since Darwin's day, there has been an accumulation of biological evidence, pointing to the conclusion that the occurrence of the same character in two forms of life—instead of being necessarily due to their "inheritance" of that character from some common ancestor who also possessed it—is often due to a tendency to parallel progression, which is

[1] Nägeli, C. von (1884), p. 522.

naturally strongest in related forms, but which may also show itself in those which are systematically remote from one another. This evidence—much of which happens to have been drawn from Monocotyledons—we must now attempt to analyse.

It was the remarkable acumen of Naudin which enabled him in 1856, as a result of his study of *Cucurbita maxima, C. Pepo,* and *C. moschata,* to enunciate the principle of parallelism in variation. After naming various points in which these species resemble one another, he adds, "Mais le trait le plus saillant de leurs analogies consiste dans des variations de même ordre et en quelque sorte parallèles: les modifications dont une espèce est susceptible se présentant presque toutes chez les deux autres[1]." To Naudin, however, this discovery was merely incidental and he does not seem to have followed it up. The first writer who can be said actually to have isolated and studied the fact that the different species of a genus, when they vary, vary along parallel lines, and not fortuitously in any and every direction, was Duval-Jouve[2], whose paper on *Variations parallèles des types congénères* has only in recent years met with full appreciation. An example will show the nature of his method; he demonstrates that, within each of the different species of *Juncus* native to France, one may find variations presenting either a compact or loose inflorescence; capsules whose colour ranges from shiny black to greenish brown; capsules varying in volume. He also records the existence of series of parallel variations within the species of many Grasses and Cyperaceae. The conclusion which Duval-Jouve draws seems to me of such importance as to warrant the citation of it in his own words:

"Par ce qui précède, on voit que lorsque des espèces congénères ont une distribution étendue, elles se modifient parallèlement les unes aux autres, ou, en d'autres termes, présentent des suites de variations ou de variétés qui se correspondent de l'une à l'autre. Or, quand on a constaté, sur la majeure partie des types d'un genre, une certaine série de variations parallèles, n'est-on pas autorisé à inférer que non-seulement ces modifications sont compatibles avec la loi essentielle de ce genre, mais qu'elles dépendent d'une loi propre et particulière à son organisation, ensuite de laquelle telles et telles parties ont une flexibilité qu'elles n'ont pas sur les autres genres? Et dès lors n'est-on pas fondé à prévoir, dans une très-grande généralité, que la même série de variations peut se reproduire sur les autres types congénères?"

Buchenau[3], who also used the Juncaceae as a text, carried Duval-Jouve's idea into wider cycles of affinity—though without, apparently, being acquainted with the earlier writer's work. He thinks that the form of inflorescence in which bracteoleless flowers are congregated into a head, is a derivative form, but that it occurs in more than one line. He also points

[1] Naudin, C. (1856).   [2] Duval-Jouve, J. (1865).   [3] Buchenau, F. (1881).

out that a single species of one sub-genus may show some feature characteristic of another sub-genus—that is to say, the same character may recur in different phyletic lines. He considers that it is impossible to ascribe this phenomenon to external agencies, and he holds that the different stocks have a common tendency to vary in certain definite directions.

Three years after the appearance of Duval-Jouve's paper on parallel variations, Darwin's great work on *The Variation of Animals and Plants under Domestication* was published. In this book he discusses "Analogous or Parallel Variations," and quotes a number of cases of parallel variation in distinct species. But he was unaware of the existence of Duval-Jouve's work, and his classification of these parallel variations shows that his grasp of their meaning was incomplete. He subdivides them under two heads, "firstly, those due to unknown causes having acted on organic beings with nearly the same constitution, and which consequently vary in an analogous manner; and secondly, those due to the reappearance of characters which were possessed by a more or less remote progenitor[1]." The precise meaning of these sentences is not altogether clear, but they suggest that Darwin, even when faced by the facts of parallel variation, could not get away from the conviction that there were only two causes which could lead variation into a definite path—the influence of the external environment, and a tendency to reversion to the characters of an ancestor. He never *fully* allowed for that spontaneous variation—due to the tendency to change, inherent in the very nature of living beings—which is not fortuitous, but follows a path conditioned by the particular organisation of the species in question. Such variation, when it occurs in related species, naturally follows parallel routes. This parallelism is not, however, confined to the species of a genus, but can, as we shall shortly see, be traced in much larger groups.

More than one systematist has drawn attention to parallel seriation in Monocotyledonous families, though without analysing the evolutionary meaning of this parallelism. Baker[2], for instance, divides the Liliaceae into three sub-families—the Liliaceae proper, with loculicidal capsules; the Colchicaceae, with septicidal capsules; and the baccate Asparagaceae. He points out that, to a certain extent, the tribes in the baccate series "run parallel with, and represent, the non-bulbous tribes in the capsular series." He considers that the Colchicaceae, in the capsular series, are represented by the Streptopeae, in the baccate series; the Hemerocallideae by the Convallarieae, etc. Baillon[3], again, traces a close parallelism between Liliaceae and Amaryllidaceae. He calls *Agave* and *Barbacenia* in the Amaryllidaceae the analogues of *Aloe*, *Yucca* and *Dracaena* in the Liliaceae; and the Dioscoreaceae, which he includes in Amaryllidaceae, the analogues of the Smilaceae.

[1] Darwin, C. (1868), vol. ii, p. 348.   [2] Baker, J. G. (1875).   [3] Baillon, H. (1894[1]).

Evidence for parallelism in the Liliiflorae may be drawn from anatomy as well as from floral morphology. As Scott and Brebner[1] pointed out thirty years ago, there is a remarkable general agreement between the mode of secondary thickening, not only in the arborescent Liliaceae, and in the narrowly limited group of the Iridaceae to which *Aristea* belongs, but also in the Dioscoreaceae. These authors concluded that this peculiar anatomical development must have been separately evolved in these three distinct groups of the Liliiflorae, and they drew special attention to it as an example of "the origination of similar, and apparently homologous structures in groups of organisms which are phylogenetically distinct."

Other writers, such as Ganong[2], E. A. N. Arber[3], Gates[4], Bower[5] and Hutchinson[6], have emphasised the part which parallelism plays in evolution, but it is to Vavilov of Petrograd that we owe a general discussion of this principle, as applied to the higher plants, which has placed the subject on a broad and satisfactory basis[7]. Vavilov gives much attention to the Gramineae, and as an example of his method we may cite his account of the variation of two meadow Grasses widely distributed over European and Asiatic Russia —*Agropyrum repens*, Beauv., and *A. cristatum*, Gaertn. A comparison of plants growing under typical wild conditions shows that both species exhibit variations which are differentiated by the following characters:

1.   Bearded or beardless ears.
2.   Hairy or smooth glumes.
3.   Yellow, red or black ears.
4.   Anthocyanin in the ear, or no anthocyanin.
5.   Spreading (lying) or erect form of seedling.
6.   Thin or thick straw.
7.   Loose or dense ears.
8.   Ears covered with wax, or without wax.
9.   Yellow or violet anthers.
10. Short or long stems.
11. Hairy or smooth leaves.
12. Narrow or broad leaves.
13. Early or late seasonal development.
14. Hydrophilous or xerophilous habit.

The only variation not known to be repeated in the two species, is that of *A. cristatum* in which the straw is full of pith, whereas all other known forms of the two species have a hollow straw; but a variety of *A. repens* in which there is pith may possibly await future discovery.

[1] Scott, D. H. and Brebner, G. (1893).   [2] Ganong, W. F. (1894).
[3] Arber, E. A. N. (1903).   [4] Gates, R. R. (1917) etc.   [5] Bower, F. O. (1918).
[6] Hutchinson, J. (1923).   [7] Vavilov, N. I. (1922).

Vavilov goes on to show that homologous variation occurs, not only in species belonging to the same genus, but also in species of related genera. He enumerates 34 alternative pairs of varietal characters, all of which are met with in Rye (*Secale cereale*, L.) and Wheat (*Triticum vulgare*, Vill.). This parallelism was so remarkable in its general completeness, that when in 1916 Vavilov discovered varieties of Wheat in which the leaves had no ligule, he predicted that corresponding varieties would be found in the Rye —and in 1918 this prediction was fulfilled by the appearance of such a variety among Pamirian Rye sown at his experimental station. He also predicted that a Rye with hairy ears would eventually come to light, to correspond with the existing hairy-eared Wheats—and here again the forecast was soon justified.

Having shown that species and genera, more or less nearly related to each other, are characterised by similar variation-series, Vavilov carries the matter farther by demonstrating that in certain cases a definite cycle of variability is traceable in all the genera of a given family.

Some years ago I entered on the study of Monocotyledonous leaves without any preconceptions about parallelism of development—having, indeed, at that time, a very inadequate knowledge of what had been written on the subject. But the principle soon became evident to me, and as my attention was turned to family after family, I began to feel that it would be difficult to exaggerate the importance of homologous variation.

In the Iridaceae[1] this parallelism is particularly conspicuous, as the range of leaf form in this family offers a field in which the recurrence of types can be clearly traced. We may illustrate it by comparing the Iridoideae and Ixioideae—the two main tribes of the family. In each of the genera *Iris*, *Sisyrinchium*, *Trimezia* and *Bobartia*—all belonging to the Iridoideae—we meet with ensiform leaves, and also leaves which are approximately radial; while in *Cipura*, *Cypella*, and the Shell-flower, *Tigridia*, which are members of the same tribe, there are foliated leaves. In the second tribe, the Ixioideae, we again meet with ensiform leaves, in such genera as the Kaffir-lily, *Schizostylis*; ensiform leaves and also leaves showing various modifications of radial structure, in *Gladiolus*; and foliated leaves in the Baboon-root, *Babiana*[2].

Parallelism in leaf development within a family may also be illustrated from the Palms, for the "fan" and "feather" leaf-types must have been evolved more than once in the development of the group, since examples of these two classes may be found among Palms which are nearly allied. The sub-family Coryphoideae, for instance, includes two tribes—the Phoeniceae, which are feather-leaved, and the Sabaleae, which are fan-leaved; while

---

[1] Illustrations of the leaves of Irids are given in figs. liv, p. 76, lxvii, p. 90, lxxvii—lxxx, pp. 102—107, xcv, p. 120, xcvi, p. 122.   [2] Arber, A. (1921[1]).

the same thing is repeated in the Lepidocaryineae, which are divided into the fan-leaved Mauriteae and the feather-leaved Metroxyleae[1].

When we take a wider survey, and, instead of considering a single family, make a comparative study of the leaves of a series of families belonging to an individual cohort, we are struck at once by the same kind of parallelism —identical in type, though affecting a larger field. We may take the Helobieae[2] as an example. This cohort consists of the marsh and aquatic families, Alismaceae, Butomaceae, Juncaginaceae, Potamogetonaceae, Naiadaceae, Aponogetonaceae, and Hydrocharitaceae. We find, on examining the leaves of this group, that there are three leaf-types which recur throughout the constituent families: firstly, a simple leaf with a sheathing base and a more or less radial limb, which is found in six of the seven families; secondly, a leaf with a sheathing base and a flat ribbon-like limb, which occurs in each of the seven families; and thirdly, a leaf with petiole and blade differentiated from one another, which is known from five of the seven families.

But comparison of the forms within a cohort by no means exhausts the data as to parallel development supplied by Monocotyledonous leaves. We may institute a comparison between the leaves of distinct cohorts, and for this purpose I propose to glance at those of the Helobieae, Liliiflorae, Farinosae and Microspermae. Four principal types of leaf recur throughout these cohorts: a simple linear leaf, with a radial limb; a simple linear leaf with an ensiform limb; a simple linear leaf, whose flat limb has one series of bundles; and a more complex type of leaf, the limb of which is differentiated into a petiole and a blade, the latter often tending to be more or less cordate at the base. I include here an illustration of the leaf-types of the Orchidaceae (Microspermae), because this family was not considered in the chapter dealing with leaf structure (fig. clx; cp. also fig. xiii, p. 27). It will be seen that the Orchids provide instances of various types which we studied in the Liliiflorae. The ensiform leaf of *Oberonia iridifolia*, Lindl. (fig. clx, 6), may, as its name implies, be compared with that of an *Iris*, and fig. 4 shows, for the case of *Dendrobium anceps*, Sw., that this resemblance is carried into the anatomy. The linear leaf of *Apostasia Wallichii*, R. Br. (fig. 3) is of a similar type to that of *Hemerocallis*, while the thick leaf of *Brassavola Martiana*, Lindl. (fig. 5) has a vascular scheme recalling that of *Dasylirion*; and the plicate limb of *Sobralia Lowii* (figs. 2 A—C) can be paralleled in *Curculigo* (fig. xcvii, 17 F, p. 123). When we turn to those leaves in which the limb is more sharply differentiated from the sheath, we find the *Funkia* type recurring in *Liparis Cathcartii*, Hook. fil. (cp. fig. xli, p. 64, with fig. clx, 7), while the cordate base of *Nephelaphyllum tenuiflorum*, Blume (fig. clx, 8) recalls the leaves of various Liliiflorae such as *Eurycles* (fig. c, p. 126).

[1] Arber, A. (1922[2]).   [2] Arber, A. (1921[2]).

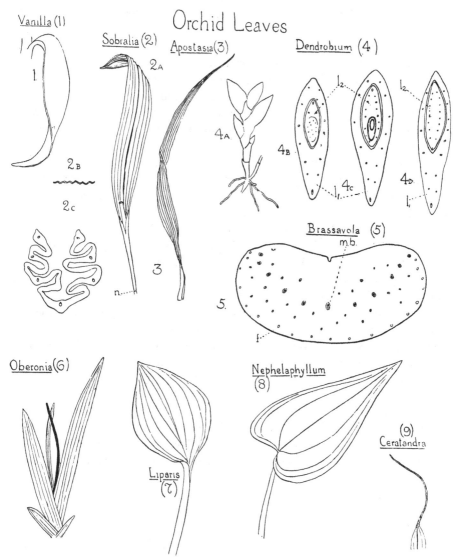

Fig. clx. Orchid leaves. Fig. 1, *Vanilla phaeantha*, Reichb., leaf (×½). Figs. 2 A—C, *Sobralia Lowii* (garden var.); fig. 2 A, leaf (×½); *n.*, node; fig. 2 B, limb cut across to show plication; fig. 2 C, transverse section of limb of young leaf (× 14). Fig. 3, *Apostasia Wallichii*, R.Br., leaf (×½). Figs. 4 A—D, *Dendrobium anceps*, Sw.; fig. 4 A, shoot (×½); figs. 4 B—D, transverse sections from a series upwards from below through an apical bud (× 14) to show base, 4 B, sheath, 4 C, limb, 4 D, of leaf *l₂* enclosed in sheath of leaf *l₁*. Fig. 5, *Brassavola Martiana*, Lindl., transverse section of limb of leaf (× 7); *m.b.*, median bundle ; *f.*, dorsal series of fibrous strands. Fig. 6, *Oberonia iridifolia*, Lindl., shoot (×½); inflorescence-axis marked in black. Fig. 7, *Liparis Cathcartii*, Hook. fil., leaf (×½). Fig. 8, *Nephelaphyllum tenuiflorum*, Blume, leaf (×½). Fig. 9, *Ceratandra* sp., leaf (×½). [A.A.]

A corresponding parallelism between two other cohorts—Glumiflorae and Farinosae—is suggested by a descriptive term used by Lindman[1]; he labels the Grasses of the S. American forests "the Commelinaceae type," on account of their characteristic leaf-form.

Comparisons such as these might be carried much farther, for the remarkable parallelism found between Monocotyledonous leaves belonging to different cycles of affinity extends even to small anatomical details. But perhaps enough has been said to show that, even if no other data were available, the fact of parallelism in evolution might be deduced from leaf structure alone.

If parallelism on different phyletic lines is, indeed, a deep-seated character, one would expect to find it in the animal as well as the vegetable kingdom. And this expectation is justified; zoologists have, moreover, obtained from the evidence of fossils, far more complete and convincing data as to progression on parallel lines, than botanists have, up to the present, discovered[2]. Parallel lineages have been studied in many groups such as Ammonites, Graptolites[3], Cretaceous Polyzoans, Palaeozoic Corals and several groups of Tertiary Mammals. The fact that this evidence is palaeontological, and that the actual succession in time of the different forms thus comes under observation, makes the study of these parallel progressions more illuminating that those which are based upon the examination of modern types alone.

When we ask ourselves what is to be the effect of the conception of parallelism upon our systems of classification, the answer is necessarily somewhat disturbing. It has become clear, for instance, that, among the Graptolites, *Tetragraptus*, *Diplograptus* and *Monograptus* are not true genera, as was supposed when they were named; they prove merely to be "corresponding stages in the evolution of several parallel groups[2]." The application of similar ideas to plants has not yet been widely made, though Bower[4], for the case of the Ferns, has recognised the necessity of interpreting taxonomic systems in the light of what he aptly calls phyletic "slide" or "drift." He points out, for instance, that the generic name *Acrostichum* is applied to Ferns which are not necessarily of common descent, but which have reached that particular stage of soral development in which freely exposed sporangia spring from a considerable area of leaf-surface. How far the idea that a genus is merely an evolutionary platform—occupied by species not necessarily of common descent—can be applied to Flowering Plants, is at present uncertain. Though many cases of polyphylesis will doubtless be found among Angiospermic genera, as at present constituted, I am inclined to think that the majority may safely be treated as having each been differentiated from a single stock; in various cases the evidence of geographical

[1] Lindman, C. A. M. (1899).   [2] Woodward, A. Smith (1923) and papers there cited.
[3] Elles, G. L. (1922) and (1923).   [4] Bower, F. O. (1915), (1918) etc.

distribution sets the seal upon this conclusion. And it seems to me, in general, that the divisions marked out by taxonomists—however our interpretation of them may vary—will continue to hold their ground; certain units, however, which were formerly set down, without hesitation, as comparable to human families, may come to be regarded rather as classes in a school—the members of which are associated, not by reason of relationship, but because they have attained to the same stage of development.

So much for the results to the systematist of the conception of parallel progressions—it remains to consider its significance as throwing light on the nature of living things. It can scarcely be doubted that this idea carries with it the implication that the direction in which any race progresses is due—not to the action of external agencies, nor to the exigencies of competition—but to the inherent nature of the race itself. In speaking of the predominant part played by inherent tendencies, I am far from wishing to deny that environment has its influence. I think, however, that, in the history of a race, environment is of no greater importance in comparison with the innate qualities of the stock, than "Nurture" in comparison with "Nature" in the history of the individual. Nurture may indeed cause remarkable changes, but these changes are confined within the inexorable limits set by Nature. It is true that Bower, in his discussion of the Ferns, appears to favour the view that phyletic drift may possibly be determined by external conditions[1]. I have, however, failed to find, either in the data which he brings forward, or in the results of my own study of Monocotyledonous leaves, any evidence for this view. It seems to me that *the tendency to progress in a certain definite direction is as much an inherent character of a given race, as are the features of its chemistry or morphology.* It is the geneticists who have cleared the ground for this idea, for their work has opened our eyes to the fact that we have not fully analysed the characters of an individual organism until we are acquainted, not only with its observable features, but also with its potentialities if bred from. In the same way a group cannot be regarded as completely understood, until we have laid bare the lines of progression which it is its destiny to follow. This point of view, though it has been somewhat overlooked in Botany, has been clearly grasped in Zoology; Smith Woodward has expressed the opinion that, from the standpoint of the palaeontologist, the "ideal definition of any category includes not merely the usual diagnostic characters, but also a statement of tendencies in evolution." That these tendencies in evolution should so often lead to parallel developments, need not surprise us. Even if we admit the specificity of protoplasm, thus assuming that the physico-chemical frame-work of the fertilized egg-cell has distinctive features in every species, these differences, within any such group as the Angiosperms, must yet be of minor importance as compared with the

[1] Bower, F. O. (1915).

resemblances. So it is natural that, when the protoplasm of related species initiates changes, these changes should follow corresponding lines; and this will apply, not only to closely related species, but to all branches of the Angiosperms, since there must be a certain similarity of plan in the protoplasm of the fertilised eggs in all members of the group, to account for the similarities between the plants to which they give rise.

Vavilov suggests a useful analogy from the domain of organic chemistry, when he points out that the series of hydrocarbons derived from every individual type-hydrocarbon, run, as it were, parallel to one another, and that, in the same way, the variations originating by protoplasmic change in a given species may be expected to form series running parallel with those initiated by other species. Salisbury, again, has illustrated convergent evolution by pointing to the many isotopes—such as the two types of lead derived from uranium and thorium respectively—which have now been shown to possess apparently identical chemical properties, associated with different atomic weights.

But the question of the particular way in which we should visualise the mechanism which lies behind the phenomena of parallel progression, is—for us, at the moment—a side issue. We have rather to ask ourselves how the conception of parallel progression affects our broad view of the organic world. To me its primary outcome seems to be, that it leaves no loophole of escape from the belief that the directions taken in the evolutionary development of any living thing are due to some compulsion inherent in its very nature, and are in no wise attributable to external causes. If we accept this view, with all its implications, we can no longer picture the evolution of plants, except as a journey along lines that are essentially fore-ordained. The secret of the great underlying principle, whatever it be, which determines these lines of progress, is entirely hidden from us. Morphological study has revealed something, and, we hope, will in the future reveal more, of the laws according to which it works, but, try as we may, we have never yet succeeded in lifting so much as a corner of the veil which hides the mystery itself. I doubt whether, even to-day, we can better the conception—foreshadowed nearly a century ago by Theodor Schwann—that not only the eternal harmony of the stars, and the changing phases of the inorganic world, but even the course of the stream of life in its passage down the ages, were determined, once and for ever, when the reign of law began, in the Dawn of All Things.

BIBLIOGRAPHY

[This bibliography is limited to the titles of memoirs actually cited in the text; it thus includes only a small proportion of the existing literature relating to Monocotyledons. References are given to the pages where each memoir is mentioned, and to any figures derived from it.]

Adamson, R. S. (1924) Preliminary Note on Secondary Growth in some Iridaceae. Nature, vol. CXIV, p. 262, Aug. 16, 1924. [p. 41.]

Arber, A. (1917) On the Occurrence of Intrafascicular Cambium in Monocotyledons. Ann. Bot., vol. XXXI, 1917, pp. 41–5, 3 text-figs. [p. 40, and Fig. xx, p. 41.]

Arber, A. (1918¹) Further Notes on Intrafascicular Cambium in Monocotyledons. Ann. Bot., vol. XXXII, 1918, pp. 87–9, 4 text-figs. [p. 40, and Figs. xviii, p. 39, xx, p. 41.]

Arber, A. (1918²) The Phyllode Theory of the Monocotyledonous Leaf, with Special Reference to Anatomical Evidence. Ann. Bot., vol. XXXII, 1918, pp. 465–501, 32 text-figs. [pp. 54, 64, 87, 103, 125, and Figs. xxxviii, p. 62, xli, p. 64, lvii, p. 78, lxxxiii, p. 110, lxxxv, p. 113.]

Arber A. (1919¹) The Vegetative Morphology of Pistia and the Lemnaceae. Proc. Roy. Soc. B, vol. XCI, 1919, pp. 96–103, 8 text-figs. [p. 148, and Figs. cxvii, p. 149, cxviii, p. 150, cxix, p. 151.]

Arber, A. (1919²) On Atavism and the Law of Irreversibility. Amer. Journ. Sci., vol. XLVIII, pp. 27–32, 1919. [p. 130.]

Arber, A. (1919³) The "Law of Loss" in Evolution. Proc. Linn. Soc. Session 131, 1918–9, pp. 70–8. [p. 130]

Arber, A. (1919⁴) Studies on Intrafascicular Cambium in Monocotyledons (III and IV). Ann. Bot., vol. XXXIII, 1919, pp. 459–65, 7 text-figs. [p. 40, and Fig. lx, p. 81.]

Arber, A. (1920¹) Leaf-base Phyllodes among the Liliaceae. Bot. Gaz., vol. LXIX, 1920, pp. 337–40, 4 text-figs. [p. 117, and Fig. lxxxvii, p. 115.]

Arber, A. (1920²) Tendrils of Smilax. Bot. Gaz., vol. LXIX, 1920, pp. 438–42, 1 plate. [pp. 64, 127, and Fig. cii, p. 128.]

Arber, A. (1920³) On the Leaf Structure of certain Liliaceae, considered in Relation to the Phyllode Theory. Ann. Bot., vol. XXXIV, 1920, pp. 447–65, 38 text-figs. [Figs. xxxix, p. 63, xliv, p. 67, li, p. 73, lii, p. 74, liii, p. 75, lv, p. 77, lvi, p. 78, lxxvi, p. 101, lxxxvi, p. 114.]

Arber, A. (1920⁴) Water Plants: A Study of Aquatic Angiosperms. Cambridge, 1920. [pp. 89, 158, 184, 193, 200, 205, 210.]

Arber, A. (1920⁵) Studies on the Binucleate Phase in the Plant-cell. Journ. Roy. Micr. Soc., 1920, pp. 1–21, 1 plate, 2 text-figs. [p. 53.]

Arber, A. (1921¹) The Leaf Structure of the Iridaceae, considered in relation to the Phyllode Theory. Ann. Bot., vol. XXXV, 1921, pp. 301–36, 66 text-figs.
[pp. 77, 118, 133, 227, and Figs. liv, p. 76, lxviii, p. 92, lxxvii, p. 102, lxxviii, p. 103, lxxix, p. 105, lxxx, p. 107, lxxxii, p. 109, xcv, p. 120, xcvi, p. 122.]

Arber, A. (1921²) Leaves of the Helobieae. Bot. Gaz., vol. LXXII, 1921, pp. 31–8, 1 plate.
[pp. 106, 228, and Fig. lxxxiv, p. 111].

Arber, A. (1921³) Leaves of certain Amaryllids. Bot. Gaz., vol. LXXII, 1921, pp. 102–5, 8 text-figs.
[Figs. xxxii, p. 56, c, p. 126.]

Arber, A. (1922¹) On the Leaf-tips of certain Monocotyledons. Linn. Soc. Journ., Bot., vol. XLV, 1922, pp. 467–76, 14 text-figs.
[Figs. lxv, p. 87, lxxxviii, p. 116, cxxxix, p. 182.]

Arber, A. (1922²) On the Development and Morphology of the Leaves of Palms. Proc. Roy. Soc., B., vol. XCIII, 1922, pp. 249–61, 7 text-figs.
[pp. 118, 228, and Figs. xlix, p. 72, l, p. 72, xc, p. 118, xci, p. 118, xcii, p. 118, xciii, p. 119, xciv, p. 119.]

Arber, A. (1922³) Studies on Intrafascicular Cambium in Monocotyledons. V. Ann. Bot., vol. XXXVI, 1922, pp. 251–6, 8 text-figs.
[Fig. lix, p. 80.]

Arber, A. (1922⁴) On the Nature of the "Blade" in certain Monocotyledonous Leaves. Ann. Bot., vol. XXXVI, 1922, pp. 329–51, 29 text-figs.
[Figs. xxxvii, p. 61, xlviii, p. 71, xcvii, p. 123, xcviii, p. 124.]

Arber, A. (1922⁵) Leaves of the Farinosae. Bot. Gaz., vol. LXXIV, 1922, pp. 80–94, 3 plates.
[pp. 64, 106, and Figs. lviii, p. 79, lxvi, p. 88, lxxxi, p. 108.]

Arber, A. (1923¹) On the "Squamulae Intravaginales" of the Helobieae. Ann. Bot., vol. XXXVII, 1923, pp. 31–41, 5 text-figs.
[p. 96, and Figs. lxx, p. 94, lxxi, p. 95, lxxii, p. 96, lxxiii, p. 97.]

Arber, A. (1923²) On the Leaf-tip Tendrils of certain Monocotyledons. Journ. Indian Bot. Soc., vol. III, no. 6, pp. 159–69, 3 plates, 1923.
[Figs. lxii, p. 84, lxiii, p. 85, lxiv, p. 86.]

Arber, A. (1923³) Leaves of the Gramineae. Bot. Gaz., vol. LXXVI, 1923, pp. 374–88, 3 plates.
[pp. 133, 163, and Figs. cv, p. 134, cxxxiv, p. 167, cxxxv, p. 169.]

Arber, A. (1924¹) *Danae, Ruscus* and *Semele*: a Morphological Study. Ann. Bot., vol. XXXVIII, 1924, pp. 229–60, 50 text-figs.
[p. 137, Figs. cvii, p. 138, cviii, p. 139, cix, p. 140, cx, p. 142, cxi, p. 143, cxii, p. 144, cxiii, p. 145, cxiv, p. 146, cxv, 147, cxvi, p. 148.]

Arber, A. (1924²) Leaves of *Triglochin*. Bot. Gaz., vol. LXXVII, 1924, pp. 50–62, 3 plates.
[pp. 80, 83, 87, 106.]

Arber, A. (1924³) *Myrsiphyllum* and *Asparagus*: a morphological study. Ann. Bot., vol. XXXVIII, 1924, pp. 635–59, 46 text-figs.
[pp. 51, 137].

Arber, A. (1925) On the "Squamulae Intravaginales" of the Alismataceae and Butomaceae. Ann. Bot., vol. XXXIX, 1925, pp. 169–73, 11 text-figs.
[p. 96.]

Arber, A. See also Robertson (Arber), A. (1906); Sargant, E. and Robertson (Arber), A. (1905); Sargant, E. and Arber, A. (1915).

Arber, E. A. N. (1903) Homoeomorphy among Fossil Plants. Geol. Mag., Dec. 4, vol. X, pp. 385–8, 1903.
[p. 226.]

Baillon, H. (1894¹) Histoire des Plantes. Monographie des Amaryllidacées, Broméliacées, et Iridacées. Paris, 1894, pp. 1–164, 106 text-figs.
[p. 225.]

Baillon, H. (1894²) Étude d'un nouvel *Aspidistra*. Bull. Mens. de la Soc. Linn. de
Paris, 7 mars, 1894, No. 143, pp. 1129–32.
[p. 219.]

Baillon, H. (1895) Histoire des Plantes. Monographie des Palmiers. Paris, 1895,
pp. 245–404, 68 text-figs.
[p. 50.]

Baker, J. G. (1875) Revision of the Genera and Species of Asparageae. Journ. Linn.
Soc., Bot., vol. XIV, 1875, pp. 508–632, 4 plates.
[pp. 212, 225.]

Bary, A. de (1884) Comparative Anatomy of the Vegetative Organs of the Phanero-
gams and Ferns. Translated by F. O. Bower and D. H. Scott.
Oxford, 1884.
[p. 37.]

Bergson, H. (1911) Creative Evolution. Translated by Arthur Mitchell. London,
1911.
[p. 219.]

Bernard, N. (1902) Études sur la Tubérisation. Rev. Gén. de Bot., vol. XIV, 1902,
pp. 1–25, 58–71, 101–19, 139–54, 269–79, 3 plates, 38 text-figs.
[pp. 29, 197, and Fig. xiv, p. 31.]

Bernard, N. (1904) Recherches expérimentales sur les Orchidées. Rev. Gén. de Bot.,
vol. XVI, 1904, pp. 405–51, 458–76, 2 plates, 8 text-figs.
[p. 29.]

Bernard, N. (1909) L'Évolution dans la Symbiose. Ann. d. sci. nat., Ser. IX, Bot.,
vol. 9, 1909, pp. 1–194, 4 plates, 28 text-figs.
[p. 29, and Fig. xiv, p. 31.]

Black, J. M. (1913) The Flowering and Fruiting of *Pectinella antarctica* (*Cymodocea
antarctica*). Trans. Roy. Soc. South Austr., vol. XXXVII, 1913,
pp. (in reprint) 1–4, 1 plate.
[p. 200, Fig. clv, p. 203.]

Blackman, F. F. The Biochemistry of Carbohydrate Production in the Higher
(1921) Plants from the Point of View of Systematic Relationship. New
Phyt., vol. XX, 1921, pp. 2–9.
[p. 220.]

Blodgett, F. H. (1900) Vegetative Reproduction and Multiplication in *Erythronium*.
Bull. Torr. Bot. Club, vol. XXVII, 1900, pp. 305–15, 3 plates.
[pp. 154, 209.]

Bouygues, H. (1902) Structure, origine et développement de certaines formes vasculaires anormales du pétiole des Dicotylédones. Actes de la Soc.
Linn. de Bordeaux, vol. LVII (Ser. 6, vol. VII), 1902, pp. 41–176, 30
text-figs.
[p. 82.]

Bower, F. O. (1915) Presid. Add. Section K., Brit. Assoc. Adv. Sci. Rep. Australia,
1915 (for 1914), pp. 560–72.
[pp. 230, 231.]

Bower, F. O. (1918) Hooker Lecture: On the Natural Classification of Plants, as exemplified in the Filicales. Journ. Linn. Soc., Bot., vol. XLIV, No.
296, 1918, pp. 107–24.
[p. 226, 230.]

Braun, A. (1851) Betrachtungen über die Erscheinung der Verjüngung in der
Natur. Leipzig, 1851, xvi + 364 pp., 3 plates.
[p. 205.]

Brooks, F. T. (1923) Some Present-day Aspects of Mycology. Trans. Brit. Myc. Soc.,
vol. IX, 1923, pp. 14–32.
[p. 219.]

Buchenau, F. (1881) Der Verbreitung der Juncaceen über die Erde. Engler's Bot.
Jahrb., vol. I, 1881, pp. 104–41.
[p. 224.]

Buchenau, F. (1896) Ein Fall von Saison-Dimorphismus in der Gattung *Triglochin*. Abh. h. v. naturwiss. Verein zu Bremen, vol. XIII, pp. 408–12, 1896. [p. 216.]

Bugnon, P. (1921[1]) La feuille chez les Graminées. Thèses...docteur ès sciences, Université de Paris, Sér. A, No. 877 ; No. d'Ordre 1691, 109 pp., 12 text-figs. [pp. 103, 133.]

Bugnon, P. (1921[2]) La théorie de la syncotylie et le cas du *Streptopus amplexifolius* D.C. La notion de phyllode appliquée à l'interprétation du cotylédon des Monocotylédones. Comptes rendus, vol. CLXXIII, 1912, pp. 660–3. [p. 158.]

Bugnon, P. (1921[3]) Quelques critiques à la théorie de la phyllorhize et, d'une façon générale, aux théories phylogéniques fondées seulement sur l'ontogénie des plantes actuelles. Bull. de la Soc. Bot. de France, vol. LXVIII (Ser. 4, vol. XXI), 1921, pp. 495–506. [p. 179.]

Bugnon, P. (1922) Sur la ramification dichotome dans les cotylédons. Comptes rendus, vol. CLXXIV, 1922, pp. 1194–6, 2 text-figs. [p. 177.]

Bugnon, P. (1923) L'organisation libéroligneuse du cotylédon des Monocotylédones expliquée grâce aux phénomènes de dichotomie cotylédonaire. Bull. de la Soc. Linn. de Normandie, Ser. 7, vol. VI, 1923, pp. 16–35. [p. 177.]

Bugnon, P. (1924) Sur les homologies de la feuille chez les Graminées. Bull. de la Soc. Bot. de France, vol. LXXI (Ser. IV, vol. XXIV), 1924, pp. 246–51. [p. 103.]

Buscalioni, L. (1920–1921) Sulla Struttura delle Asparagacee (prelim. note). Bollettino d. Accad. Gioenia di Sci. Nat. in Catania, Fasc. 49. 1920–1. [p. 117.]

Buscalioni, L. and Lopriore, G. (1910) Il pleroma tubuloso, l' endodermide midollare, la frammentazione desnica e la schizorrizia nelle radici della *Phoenix dactylifera* L. Atti d. Accad. Gioenia, Ser. V, vol. III, 1910, 101 pp., 13 plates. [p. 25.]

Buscalioni, L. and Muscatello, G. (1908) Fillodi e Fillodopodi. Studio sulle Leguminose australiane. Atti d. Accad. Gioenia di Sci. Nat. in Catania, Ser. 5, vol. I, mem. VI, 1908, 30 pp., 4 plates. [p. 103.]

Buscalioni, L. See also **Pirotta, R. and Buscalioni, L.** (1898).

Candolle, A. P. de (1827) Organographie végétale. 2 vols. Paris, 1827. [pp. 1, 3, 6, 100, 175, 215.]

Candolle, C. de (1913) Les Ligules du *Trithrinax campestris* Drude et Grisebach. Bull. de la Soc. de Bot. de Genève, Ser. II, vol. V, 1913, No. 3, pp. 106–7, 1 plate. [p. 119.]

Chapman, R. E. (1924) The Carbohydrate Enzymes of certain Monocotyledons. Proc. Linn. Soc. Lond., 136th Sess., 1924, pp. 5, 6. [p. 215.]

Chauveaud, G. (1911) L'appareil conducteur des plantes vasculaires et les phases principales de son évolution. Ann. d. sci. nat., Bot., Ser. IX, vol. XIII, 1911, pp. 113–438, 218 text-figs. [p. 175, Figs. cxxxvi, p. 171, cxxxvii, p. 173.]

Chauveaud, G. (1914) La constitution et l'évolution morphologique du corps chez les plantes vasculaires. Comptes rendus, vol. CLVIII, 1914, pp. 343–6, 8 text-figs. [p. 11.]

Church, A. H. (1908) Types of Floral Mechanism. Part I, Types i–xii, Oxford, 1908. [p. 189.]

Church, A. H. (1919) Thalassiophyta and the Subaerial Transmigration. Bot. Memoirs, No. 3, Oxford, 1919. [p. 6.]

Church, M. B. (1919) The Development and Structure of the Bulb in *Cooperia Drummondii*. Bull. Torrey Bot. Club, vol. XLVI, 1919, pp. 337–62, 3 plates. [pp. 21, 45, Figs. xxiv, p. 47, xxv, p. 49.]

Clos, D. (1875) Des éléments morphologiques de la feuille chez les Monocotylés. Mém. de l'Acad. d. Sci. de Toulouse, Ser. VII, vol. VII, 1875, pp. 305–24. [p. 91.]

Colomb, G. (1887) Recherches sur les stipules. Thèses prés. à la faculté des sci. de Paris, Sér. A, No. 94; No. d'Ordre 595, Paris, 1887, 76 pp., 47 text-figs. [p. 98.]

Compton, R. H. (1912) An Investigation of the Seedling Structure in the Leguminosae. Journ. Linn. Soc. (Bot.), vol. XLI, 1912–3, No. 279, 1912, pp. 1–122, 9 plates. [p. 177, Fig. cxxxvi, p. 171.]

Curtis, K. M. (1917) The Anatomy of the Six Epiphytic Species of the New Zealand Orchidaceae. Ann. Bot., vol. XXXI, 1917, pp. 133–49, 6 plates. [pp. 16, 89.]

Dalitzsch, M. (1886) Beiträge zur Kenntniss der Blattanatomie der Aroideen. Bot. Centralbl., vol. XXV, 1886, pp. 153–6, 184–6, 217–9, 249–53, 280–5, 312–8, 343–9, 1 plate. [p. 212.]

Darwin, C. (1868) The Variation of Animals and Plants under Domestication. 2 vols., London, 1868. [p. 225.]

Darwin, C. (1904) The Various Contrivances by which Orchids are Fertilised by Insects. 2nd edn., London, 1904. [p. 195.]

Deinega, V. (1898) Beiträge zur Kenntniss der Entwickelungsgeschichte des Blattes und der Anlage der Gefässbundel. Flora, vol. LXXXV, 1898, pp. 439–98, 1 plate, 22 text-figs. [pp. 73, 82.]

Delpino, F. (1896) Applicazione di nuovi criterii per la classificazione delle piante (Monocotiledoni). Mem. d. R. Accad. d. Scienze dell' Istituto di Bologna, Ser. V, vol. VI, 1896–7, pp. 84–116. [p. 217.]

Depéret, C. (1909) The Transformations of the Animal World, London, 1909. (Translation of Les Transformations du Monde Animal, Paris, 1907.) [pp. 218, 219.]

Domin, K. (1909) Morphologische und phylogenetische Studien über die Familie der Umbelliferen. Bull. Int. Acad. Sci. Prague. Teil I, vol. XIII (1909 for 1908), pp. 108–53, 3 plates, Teil II, vol. XIV (1909), pp. 1–52, 2 plates, 10 text-figs. [p. 100.]

Döring, E. (1910) Das Leben der Tulpe. Sondershausen, 1910. [p. 150.]

Drabble, E. (1904) On the Anatomy of Roots of Palms. Trans. Linn. Soc. Lond., ser. II, vol. VI, 1901–5, pt. 10, 1904, pp. 427–90, 4 plates, 22 text-figs. [p. 25.]

Drabble, E. (1906) The Transition from Stem to Root in some Palm Seedlings. New Phyt., vol. V, 1906, pp. 56–66, 7 text-figs. [p. 87.]

Duval-Jouve, J. (1865)   Variations parallèles des types congénères. Bull. Soc. Bot. de France, vol. XII, 1865, pp. 196-211.
[p. 224.]

Duval-Jouve, J. (1877)   Étude histotaxique des cladodes du *Ruscus aculeatus* L. Bull. Soc. Bot. de France, vol. XXIV, 1877, pp. 143-8.
[p. 131, 146, 179.]

Dykes, W. R. (1913)   The Genus *Iris*. Cambridge, 1913.
[p. 209.]

Elles, G. L. (1922)   The Graptolite Faunas of the British Isles. A Study in Evolution. Proc. Geol. Assoc., vol. XXXIII, 1922, pp. 168-200, 15 text-figs. and a table.
[p. 230.]

Elles, G. L. (1923)   Evolutional Palaeontology in Relation to the Lower Palaeozoic Rocks, Presid. Add. Sect. C., Brit. Assoc. Adv. Sci., Rep. Liverpool, 1923 (for 1922), p. 83-107, 5 text-figs.
[p. 230.]

Engler, A. (1877)   Vergleichende Untersuchungen über die morphologischen Verhältnisse der Araceae. II Theil. Ueber Blattstellung und Sprossverhältnisse der Araceae. Verhandl. der Ksl. Leop.-Carol.-Deutschen Akad. der Naturforscher, vol. XXXIX, No. 4, 1877, pp. 159-232, 6 plates.
[p. 149.]

Engler, A. (1881-1884)   Beiträge zur Kenntniss der Araceae, I-V. Engler's Bot. Jahrb., vol. I, 1881, pp. 179-90, 480-8; vol. IV, 1883, pp. 59-66, 341-52, 1 plate; vol. V, 1884, pp. 141-88, 287-336, 5 plates.
[pp. 43, 70, 189, 212.]

Engler, A. (1892)   Die systematische Anordnung der monokotyledoneen Angiospermen. Abhandl. d. k. Akad. d. Wiss. Berlin, 1892, Abh. II, 55 pp.
[p. 220.]

Ewing, J.   See Priestley, J. H., and Ewing, J. (1923).

Falkenberg, P. (1876)   Vergleichende Untersuchungen über den Bau der Vegetationsorgane der Monocotyledonen. Stuttgart, 1876.
[p. 37.]

Farmer, J. B. (1918)   On the Quantitative Differences in the Water-Conductivity of the Wood in Trees and Shrubs. Part I—The Evergreens. Part II—The Deciduous Plants. Proc. Roy. Soc. B, vol. XC, 1918, pp. 218-50, 3 text-figs.
[p. 37.]

Fraine, E. de (1910)   The Seedling Structure of certain Cactaceae. Ann. Bot., vol. XXIV, 1910, pp. 125-75, 37 text-figs.
[Fig. cxxxvi, p. 171.]

Fraine, E. de   See also Hill, T. G., and Fraine, E. de (1908-1910).

Fries, R. E. (1911)   Ein unbeachtet gebliebenes Monokotyledonenmerkmal bei einigen Polycarpicae. Ber. d. deutsch. Bot. Gesell., vol. XXIX, 1911, pp. 292-301, 6 text-figs.
[p. 131.]

Fritsch, K. (1905)   Die Stellung der Monokotylen im Pflanzensystem. Bot. Jahrb., vol. XXXIV, 1905, Beiblatt 79, pp. 22-40.
[p. 217.]

Fyson, P. F. (1921)   The Indian Species of *Eriocaulon*. (Part I.) Journ. of Indian Botany, vol. II, 1921, pp. 133-50.
[p. 211.]

Gaisberg, E. von (1922)   Zur Deutung der Monokotylenblätter als Phyllodien, unter besonderer Berücksichtigung der Arbeit von A. Arber; "The Phyllode Theory of the Monocotyledonous Leaf, with Special Reference to Anatomical Evidence." Flora, N.F., vol. XV, 1922, pp. 177-90, 3 plates.
[pp. 100, 103.]

Ganong, W. F. (1894) Beiträge zur Kenntniss der Morphologie und Biologie der Cacteen. Flora, vol. LXXIX (Ergänz.), 1894, pp. 49–86, 17 text-figs. [p. 226.]

Garlick, C. (1918) Note on the Sacred Tree in Mesopotamia. Proc. Soc. Bibl. Arch., vol. XL, 1918, pp. 111–2, 1 plate. [p. 192.]

Gates, R. R. (1917) A Systematic Study of the North American genus *Trillium*, Its Variability, and Its Relation to *Paris* and *Medeola*. Ann. Missouri Bot. Garden, vol. IV, 1917, pp. 43–92, 3 plates, 1 map. [pp. 211, 226.]

Gates, R. R. (1918) A Systematic Analytical Study of certain North American Convallariaceae, considered in regard to their Origin through Discontinuous Variation. Ann. Bot., vol. XXXII, 1918, pp. 253–7. [p. 210.]

Gatin, V. C. (1920) Recherches anatomiques sur le pédoncule et la fleur des Liliacées. Rev. Gén. de Bot., vol. XXXII, 1920, pp. 369–437, 460–528, 561–91, 60 text-figs. [pp. 8, 210.]

Ghose, S. L. (1923) An Example of Leaf-enation in *Allium ursinum* L. New Phyt., vol. XXII, 1923, pp. 49–58, 10 text-figs. [p. 103.]

Glück, H. (1901) Die Stipulargebilde der Monokotyledonen. Verhandl. d. Naturhist.-Med. Vereins zu Heidelberg. N.F., vol. VII, Heft 1, 1901, pp. 1–96, 5 plates, 1 text-fig. [p. 98.]

Glück, H. (1919) Blatt- und blütenmorphologische Studien. Jena, 1919, 696 pp., 7 plates, 284 text-figs. [pp. 103, 135.]

Goebel, K. (1922[1]) Organographie der Pflanzen. Teil III, Heft 1. 2nd edn. Jena, 1922. [p. 103.]

Goebel, K. (1922[2]) Erdwurzeln mit Velamen. Flora, N.F., vol. XV (G.R. vol. CXV), Heft 1, 1922, pp. 1–26, 3 text-figs. [p. 16.]

Goethe, J. W. von (1790) Versuch die Metamorphose der Pflanzen zu erklären. Gotha, 1790. [p. 3, 179.]

Grassmann, P. (1884) Die Septaldrüsen. Flora, Jahrg. LXVII, 1884, pp. 113–36, 2 plates. [p. 194.]

Gravis, A. (1898) Recherches anatomiques et physiologiques sur le *Tradescantia virginica* L. Mémoires couronnés et Mém. de savants étr. Acad. roy. des sci., des lettres et des beaux-arts de Belgique, vol. LVII, 1898, 304 pp., 27 plates. [p. 131, and Fig. civ, p. 132.]

Groom, P. (1895) Contributions to the Knowledge of Monocotyledonous Saprophytes. Journ. Linn. Soc., Bot., vol. XXXI, 1895–7, No. 214, 1895, pp. 149–215, 3 plates, 3 text-figs. [p. 48.]

Guillaud, M. (1924) Sur la préfeuille des Graminées. Bull. de la Soc. Linn. de Normandie, Ser. VII, vol. VII, 1924, pp. 41–99, 52 text-figs. [p. 133.]

Guppy, H. B. (1919) Plant-Distribution from the Standpoint of an Idealist. Journ. Linn. Soc., Bot., vol. XLIV, 1919, pp. 439–72. [pp. 3, 218.]

Haberlandt, G. (1914) Physiological Plant Anatomy, translated from the fourth German edition by Montagu Drummond. London, 1914. [pp. 16, 37.]

Hallier, H. (1912)   L'Origine et le système phylétique des Angiospermes exposés à l'aide de leur arbre généalogique. Archives Néerland. Ser. III B, vol. I, 1912, pp. 146–234, 7 text-figs., 6 tables.
[p. 114.]

Hauman, L. (1916)   Les Dioscoréacées de l'Argentine. Anales del Museo Nac. de Hist. Nat. de Buenos Aires, vol. XXVII, 1916, pp. 441–513, 33 text-figs.
[p. 130.]

Heim, F. (1896)   Un substitutif possible du Safran. Bull. Mens. Soc. Linn. de Paris. 6 Mars 1896, No. 155, pp. 1231–2.
[p. 221].

Henslow, G. (1911)   The Origin of Monocotyledons from Dicotyledons, through Self-adaptation to a Moist or Aquatic Habit. Ann. Bot., vol. XXV, Part II, 1911, pp. 717–44.
[p. 103.]

Hill, A. W. (1906)   The Morphology and Seedling Structure of the Geophilous Species of *Peperomia*, together with some Views on the Origin of Monocotyledons. Ann. Bot., vol. XX, 1906, pp. 395–427, 2 plates, 3 text-figs.
[p. 178.]

Hill, T. G. and Fraine, E. de (1908–1910)   On the Seedling Structure of Gymnosperms. I, Ann. Bot., vol. XXII, 1908, pp. 689–712, 1 plate, 8 text-figs.; II, Ibid., vol. XXIII, 1909, pp. 189–227, 1 plate, 11 text-figs.; III, Ibid., vol. XXIII, 1909, pp. 433–58, 1 plate, 4 text-figs.; IV, Ibid., vol. XXIV, 1910 pp. 319–33, 2 plates, 3 text-figs.
[Fig. cxxxvi, p. 171.]

Holm, T. (1904)   The root-structure of North American terrestrial Orchideae. Amer. Journ Sci., Ser. IV, vol. XVIII, 1904, pp. 197–212, 4 text-figs.
[p. 16.]

Hutchinson, J. (1923)   Contributions towards a Phylogenetic Classification of Flowering Plants. I. Bull. Misc. Information, Royal Bot. Gard., Kew, No. 2, 1923, pp. 65–89, 1 map, 1 diagram.
[p. 226.]

Irmisch, T. (1850)   Zur Morphologie der monokotylischen Knollen- und Zwiebel-gewächse. Berlin, 1850.
(NOTE: Irmisch's works cannot be cited in full here. The titles of those relating to aquatic Monocotyledons will be found in Arber, A. (1920⁴); for other references see the Royal Society's Catalogue.)
[pp. 47, 150, 154, 155.]

Jefferies, T. A. (1916)   The Vegetative Anatomy of *Molinia caerulea*, the Purple Heath Grass. New Phyt., vol. XV, 1916, pp. 49–71, 9 text-figs.
[p. 44.]

Jeffrey, E. C. (1917)   The Anatomy of Woody Plants. Chicago, 1917.
[p. 39.]

Karsten, H. (1847)   Die Vegetationsorgane der Palmen. Abhandl. d. k. Akad. d. Wiss., Berlin, 1849 (for 1847), pp. 73–236, 9 plates.
[pp. 50, 160.]

Kawamura, S. (1911)   On the cause of the flowering of Bamboos. Bot. Mag., Tokyo, vol. XXV, 1911, Nos. 294–6 (in Japanese; German abstract in Zeitschrift f. Pflanzenzüchtung, vol. I, p. 96, 1913).
[p. 206, 207.]

Keith, A. (1919)   The Differentiation of Mankind into Racial Types. Presid. Add. Sect. H., Rep. Brit. Assoc. Adv. Sci., Bournemouth, 1920 (for 1919), pp. 275–81.
[p. 222.]

Knudson, L. (1922) Nonsymbiotic Germination of Orchid Seeds. Bot. Gaz., vol. LXXIII, 1922, pp. 1–25, 3 text-figs.
[p. 221.]

Knudson, L. (1924) Further Observations on Nonsymbiotic Germination of Orchid Seeds. Bot. Gaz., vol. LXXVII, 1924, pp. 212–9, 3 text-figs.
[p. 221.]

Kroemer, K. (1903) Wurzelhaut, Hypodermis und Endodermis der Angiospermenwurzel. Bibl. Bot., vol. XII, 1903–4, Heft 59, 1903, 151 pp., 6 plates.
[p. 16.]

Laurent, M. (1904) Recherches sur le développement des Joncées. Ann. d. sci. nat., Bot., Ser. 8, vol. XIX, 1904, pp. 97–194, 8 plates, 16 text-figs.
[pp. 187, 199, and Fig. cliii, p. 201.]

Lee, E. (1912) Observations on the Seedling Anatomy of Certain Sympetalae. I. Tubiflorae. Ann. Bot., vol. XXVI, 1912, pp. 727-46, 1 plate, 8 text-figs.
[Fig. cxxxvi, p. 171.]

Leeuwen, W. D. van (1921) Ueber Infloreszenz-Bulbillen in der Zingiberaceen-Gattung: *Globba*. Ann. du Jardin Bot. de Buit., vol. XXI, 1921, pp. 1–17, 4 plates.
[p. 48.]

Lenoir, M. (1920) Évolution du tissu vasculaire chez quelques plantules de Dicotylédones. Ann. d. sci. nat., Bot., Ser. 10, vol. II, 1920, pp. 1–123, 91 text-figs.
[p. 176.]

Lindinger, L. (1906) Zur Anatomie und Biologie der Monokotylenwurzel. Beih. zum Bot. Centralbl., vol. XIX, Abt. I, 1906, pp. 321–58, 30 text-figs.
[p. 25.]

Lindinger, L. (1907) Ueber den morphologischen Wert der an Wurzeln entstehenden Knollen einiger *Dioscorea*-Arten. Beih. z. Bot. Centralbl., vol. XXI, Abt. I, 1907, pp. 311–24.
[pp. 24, 25.]

Lindinger, L. (1908) Die Bewurzelungsverhältnisse grosser Monokotylenformen und ihre Bedeutung für den Gärtner. Gartenflora, vol. LVII, 1908, pp. 281–91, 308–18, 367–78, 12 text-figs.
[pp. 11, 12, 31, 41, 48.]

Lindinger, L. (1909) Jahresringe bei den Monokotylen der Drachenbaumform. Naturwiss. Wochenschrift, N.F., vol. VIII, 1909, pp. 491–4, 3 text-figs.
[p. 43.]

Lindman, C. A. M. (1899) Zur Morphologie und Biologie einiger Blätter und belaubter Sprosse. Bihang till k. Svenska Vet.-Akad. Handl., vol. XXV, Afd. III, No. 4, 63 pp., 20 text-figs.
[pp. 69, 230.]

Lister, G. (1920) On Some Water Plants: A Presidential Address. Essex Nat., vol. XIX, 1918–21, Part II, 1920, pp. 103–15, 7 text-figs.
[Figs. xl, p. 64, ciii, p. 129, cxli, p. 184, cxlii, p. 185.]

Lonay, H. (1902) Recherches anatomiques sur les feuilles de l'*Ornithogalum caudatum* Ait. Mém. de la Soc. Roy. des Sci. de Liége, Ser. III, vol. IV, No. 9, 1902, pp. 1–82, 5 plates.
[pp. 51, 82, 133, and Fig. civ, p. 132.]

Lopriore, G. See **Buscalioni, L. and Lopriore, G. (1910)**.

Lotsy, J. P. (1911) Vorträge über botanische Stammesgeschichte. Jena, 1911, vol. III, Pt. I.
[p. 217.]

Lotsy, J. P. (1916) Evolution by Means of Hybridization. The Hague, 1916.
[p. 217.]

MacDougal, D. T., Richards, H. M. and Spoehr, H. A. (1919)   Basis of Succulence in Plants. Bot. Gaz., vol. LXVII, 1919, pp. 405–16. [p. 221.]

Magrou, J. (1921)   Symbiose et Tubérisation. Ann. d. sci. nat., Ser. 10, vol. III, No. 4, 1921, pp. 181–296, 9 plates, 9 text-figs. [p. 29.]

Mann, A. G. (1921)   Observations on the Interruption of the Endodermis in a Secondarily Thickened Root of *Dracaena fruticosa*, Koch. Proc. Roy. Soc. Edinb., vol. XLI, Pt. I, pp. 50–9, 13 text-figs. [p. 25.]

Martius, C. F. P. von (1835)   Die Eriocauleae, als selbständige Pflanzen-Familie aufgestellt und erläutert. Nova Acta Acad. Caes. Leop. Carol. Nat. Cur., vol. XVII, pt. I, 1835, 72 pp., 5 plates. [p. 104.]

Martius, C. F. P. von (1840 etc.)   Flora Brasiliensis. Munich, 1840 onwards. [p. 66.]

Menz, G. (1922)   Osservazioni sull' anatomia degli organe vegetativi delle specie italiane del genere *Allium* (Tourn.) L. appartenenti alla sezione "*Molium*" G. Don. Bull. dell' Ist. Bot. della R. Univ. di Sassari, vol. I, Mem. V, Feb. 1922, 27 pp., 2 plates, 7 text-figs. [pp. 103, 104.]

Meyer, A. (1885)   Ueber die Assimilationsproducte der Laubblätter angiospermer Pflanzen. § 4. Bot. Zeit., vol. XLIII, 1885, pp. 449–54. [p. 220.]

Möbius, M. (1887)   Ueber das Vorkommen concentrischer Gefässbündel mit centralem Phloëm und peripherischem Xylem. Ber. d. deutschen Bot. Gesellsch., vol. V, 1887, pp. 2–24, 2 plates. [p. 39.]

Mohl, H. von (1845)   Untersuchungen über den Mittelstock von *Tamus Elephantipes* L. Vermischte Schriften, XII, pp. 186–94. Tübingen, 1845. [p. 42.]

Mohl, H. von (1858)   Ueber die Cambiumschicht des Stammes der Phanerogamen und ihr Verhältniss zum Dickenwachsthum desselben. Bot. Zeit., vol. XVI, 1858, pp. 185-90, 193-8. [p. 35.]

Morris, D. (1886)   *Chusquea abietifolia*. Gard. Chron., 1886, vol. XXVI, p. 524. [p. 207.]

Moss, C. E. (1923)   On the presence of velaminous roots in terrestrial Orchids. Proc. Linn. Soc., 135th Session, 1923, p. 47. [p. 16.]

Müller, C. (1909)   Beiträge zur vergleichenden Anatomie der Blätter der Gattung *Agave* und ihrer Verwertung für die Unterscheidung der Arten. Bot. Zeit., Jahrg. 67, 1909, pp. 93-139, 22 text-figs., 2 plates. [p. 89.]

Muscatello, G.   See **Buscalioni, L. and Muscatello, G.** (1910).

Nägeli, C. von (1884)   Mechanisch-physiologische Theorie der Abstammungslehre. Munich and Leipsic, 1884. [p. 223.]

Naudin, C. (1856)   Nouvelles recherches sur les caractères spécifiques et les variétés des plantes du genre *Cucurbita*. Ann. d. sci. nat., Ser. IV, vol. VI, 1856, pp. 5-73, 3 plates. [p. 224.]

Onslow, M. Wheldale (1911)   The Chemical Differentiation of Species. Biochem. Journ., vol. V, 1911, pp. 445-56. [pp. 221, 222.]

Orr, M. Y. (1923)   The Leaf Glands of *Dioscorea macroura*, Harms. Notes R. B. G. Edinb., vol. XIV, 1923, pp. 57-72, 2 plates, 3 text-figs. [p. 65.]

BIBLIOGRAPHY 243

Ostenfeld, C. H. (1916)   Contributions to West Australian Botany. Part I. Introduction. The Sea-grasses of West Australia. Dansk. Bot. Arkiv, vol. II, 1916, No. 6, 44 pp., 31 text-figs.
[p. 201.]

Pallis, M. (1916)   The Structure and History of Plav: the Floating Fen of the Delta of the Danube. Journ. Linn. Soc. Bot., vol. XLIII, 1916, pp. 233–90, 15 plates, 1 text-figure.
[p. 205.]

Parkin, J. (1898)   On some points in the Histology of Monocotyledons. Ann. Bot., vol. XII, 1898, pp. 147–54, 1 plate.
[p. 56.]

Pirotta, R. and Buscalioni, L. (1898)   Sulla presenza di elementi vascolari multinucleati nelle Dioscoreacee. Annuario del R. Istituto Bot. di Roma. Anno VII, 1898; pp. 237–54, 4 plates.
[p. 39].

Porsch, O. (1910)   Die ornithophilen Anpassungen von *Antholyza bicolor* Gasp. Verhandl. d. naturf. Vereines Brünn, vol. XLIX, 1910, pp. 111–21, 2 plates, 1 text-fig.
[p. 193.]

Priestley, J. H. and Ewing, J. (1923)   Physiological Studies in Plant Anatomy. VI. Etiolation. New Phyt., vol. XXII, 1923, pp. 30–44.
[p. 11.]

Priestley, J. H. and Woffenden, L. M. (1922)   Physiological Studies in Plant Anatomy. V. Causal Factors in Cork Formation. New Phyt., vol. XXI, 1922, pp. 252–68, 1 text fig.
[p. 25.]

Queva, C. (1894)   Recherches sur l'Anatomie de l'Appareil végétatif des Taccacées et des Dioscorées. Mém. de la Soc. de Sciences de Lille, 1894, 457 pp., 18 plates.
[p. 42.]

Queva, C. (1907)   Contributions à l'anatomie des Monocotylédonées. II. Les Uvulariées rhizomateuses. Beih. zum Bot. Centralbl., vol. XXII, Abt. 2, 1907, pp. 30–77, 49 text-figs.
[p. 59.]

Ramsbottom, J. (1922)   Orchid Mycorrhiza. Trans. Brit. Myc. Soc., vol. VIII, Pts. 1 and 2, Dec. 1922, 61 pp., 6 plates.
[p. 29.]

Rendle, A. B. (1904)   The Classification of Flowering Plants. Vol. I, Gymnosperms and Monocotyledons. Cambridge, 1904.
[pp. 180, 212, 217.]

Richards, H. M.   See MacDougal, D. T., Richards, H. M. and Spoehr, H. A. (1919).

Riede, W. (1920)   Untersuchungen über Wasserpflanzen. Flora, N.F., vol. XIV, 1921, Heft 1, 1920, pp. 1–118, 3 text-figs.
[p. 66.]

Rimbach, A. (1895[1])   Jahresperiode tropisch-andiner Zwiebelpflanzen. Ber. d. deutsch. Bot. Gesellsch., vol. XIII, 1895, pp. 88–93.
[p. 19.]

Rimbach, A. (1895[2])   Zur Biologie der Pflanzen mit unterirdischem Sprosse. Ber. d. deutsch. Bot. Gesellsch., vol. XIII, 1895, pp. 141–55, 1 plate.
[p. 19, and Fig. vii, p. 19.].

Rimbach, A. (1896[1])   Ueber die Tieflage unterirdisch ausdauernder Pflanzen. Ber. d. deutsch. Bot. Gesellsch., vol. XIV, 1896, pp. 164–8.
[p. 19.]

Rimbach, A. (1896[2])   Zur Kenntniss von *Stenomesson aurantiacum* Herb. Ber. d. deutsch. Bot. Gesellsch., vol. XIV, 1896, pp. 372–4.
[pp. 19, 47.]

16—2

Rimbach, A. (1897¹)   Ueber die Lebensweise der geophilen Pflanzen. Ber. d. deutsch. Bot. Gesellsch., vol. XV, 1897, pp. 92–100.
[p. 19.]

Rimbach, A. (1897²)   Ueber die Lebensweise des *Arum maculatum*. Ber. d. deutsch. Bot. Gesellsch., vol. XV, 1897, pp. 178–82, 1 plate.
[pp. 19, 21, 209.]

Rimbach, A. (1897³)   Lebensverhältnisse des *Allium ursinum*. Ber. d. deutsch. Bot. Gesellsch., vol. XV, 1897, pp. 248–52, 1 plate.
[p. 19.]

Rimbach, A. (1897⁴)   Biologische Beobachtungen an *Colchicum auctumnale*. Ber. d. deutsch. Bot. Gesellsch., vol. XV, 1897, pp. 298–302, 1 plate.
[pp. 19, 45.]

Rimbach, A. (1898¹)   Die kontraktilen Wurzeln und ihre Thätigkeit. Fünfstück's Beitr. z. Wiss. Bot., vol. II, 1898, pp. 1–28, 2 plates.
[p. 19.]

Rimbach, A. (1898²)   Ueber *Lilium Martagon*. Ber. d. deutsch. Bot. Gesellsch., vol. XVI, 1898, pp. 104–10, 1 plate.
[p. 19.]

Rimbach, A. (1900)   Physiological observations on some perennial herbs. Bot. Gaz., vol. XXX, 1900, pp. 171–88, 1 plate.
[p. 19.]

Rimbach, A. (1902)   Physiological observations on the subterranean organs of some Californian Liliaceae. Bot. Gaz., vol. XXXIII, 1902, pp. 401–20, 1 plate.
[p. 19.]

Robertson (Arber), A. (1906)   The "Droppers" of *Tulipa* and *Erythronium*. Ann. Bot., vol. XX, pp. 429–40, 2 plates, 1906.
[pp. 150, 153, and Fig. cxx, p. 152.]

Ross, H. (1883)   Beiträge zur Anatomie abnormer Monocotylenwurzeln (Musaceen, Bambusaceen). Ber. d. deutsch. Bot. Gesellsch., vol. I, 1883, pp. 331–7, 1 plate.
[p. 15.]

Rüter, E. (1918)   Ueber Vorblattbildung bei Monokotylen. Flora, N.F., vol. X (G.R. vol. CX), 1918, pp. 193–261, 198 text-figs.
[p. 135.]

Salisbury, E. J. (1913)   The Determining Factors in Petiolar Structure. New Phyt., vol. XII, 1913, pp. 281–9.
[p. 5.]

Salisbury, E. J. (1921)   The Study of Human Implements as an Aid to the Appreciation of Principles of Evolution and Classification. New Phyt., vol. XX, 1921, pp. 179–84.
[p. 232.]

Sargant, E. (1902)   The Origin of the Seed-leaf in Monocotyledons. New Phyt., vol. I, 1902, pp. 107–13, 1 plate.
[p. 170, and Fig. cxxxvii, p. 173.]

Sargant, E. (1903)   A Theory of the Origin of Monocotyledons, founded on the Structure of their Seedlings. Ann. Bot., vol. XVII, 1903, pp. 1–92, 7 plates, 10 text-figures.
[pp. 168, 170, and Fig. cxxxvii, p. 173.]

Sargant, E. (1904)   The Evolution of Monocotyledons. Bot. Gaz., vol. XXXVII, 1904 pp. 325–45, 6 text-figs.
[p. 170.]

Sargant, E. (1908)   The Reconstruction of a Race of Primitive Angiosperms. Ann. Bot., vol. XXII, 1908, pp. 121–86, 21 text-figs.
[p. 170.]

Sargant, E. (1914)   Development of Botanical Embryology since 1870. Pres. Add. Sect. K., Rep. Brit. Ass. Adv. Sci., Birmingham, 1914 (for 1913), pp. 692–705.
[p. 156.]

Sargant, E. and Robertson (Arber), A. (1905) The Anatomy of the Scutellum in *Zea Maïs*. Ann. Bot., vol. XIX, 1905, pp. 115–23, 2 plates. [p. 163, and Fig. cxxxii, p. 165.]

Sargant, E. and Arber, A. (1915) The Comparative Morphology of the Embryo and Seedling in the Gramineae. Ann. Bot., vol. XXIX, 1915, pp. 161–222, 2 plates, 35 text-figs. [p. 163, and Figs. cxxvi, p. 161, cxxvii, p. 161, cxxx, p. 164, cxxxi, p. 164, cxxxiii, p. 166.]

Sargant, E. See also Scott, R. (Mrs D. H.) and Sargant, E. (1898).

Saunders, E. R. (1890) On the Structure and Function of the Septal Glands in *Kniphofia*. Ann. Bot., vol. V, 1890, pp. 11–25, 1 plate. [p. 194.]

Saunders, E. R. (1922) The Leaf-skin Theory of the Stem. Ann. Bot., vol. XXXVI, pp. 135–65, 34 text-figs. [pp. 34, 91, 166, 189.]

Saunders, E. R. (1923) A Reversionary Character in the Stock (*Matthiola incana*) and its Significance in regard to the Structure and Evolution of the Gynoecium in the Rhoeadales, the Orchidaceae, and other Families. Ann. Bot., vol. XXXVII, 1923, pp. 451–82, 60 text-figs. [pp. 184, 197, 199, 215, and Fig. cli, p. 198.]

Schniewind-Thies, J. (1897) Beiträge zur Kenntnis der Septalnectarien. Jena, 1897, 87 pp., 12 plates. [p. 194.]

Schoute, J. C. (1903) Die Stammesbildung der Monokotylen. Flora, vol. XCII, 1903, pp. 32–48, 1 plate. [pp. 35, 43.]

Schoute, J. C. (1907) Ueber die Verdickungsweise des Stammes von *Pandanus*. Ann. du Jardin Botanique de Buitenzorg, Ser. 2, vol. VI, 1907, pp. 115–37, 4 plates. [p. 205.]

Schoute, J. C. (1915) Sur la fissure médiane de la gaine foliaire de quelques Palmiers. Ann. du Jard. Bot. de Buit., Ser. 2, vol. XIV, pp. 57–82, 6 text-figs., 3 plates. [p. 77.]

Schoute, J. C. (1922) On Whorled Phyllotaxis. I. Growth Whorls. Recueil des travaux botaniques néerlandais, vol. XIX, 1922, pp. 184–206. [p. 54.]

Scott, D. H. (1897) On two new instances of Spinous Roots. Ann. Bot., vol. XI, 1897, pp. 327–32, 2 plates. [p. 32.]

Scott, D. H. (1920) The Relation of the Seed Plants to the Higher Cryptogams. Rep. Brit. Ass. Adv. Sci., Bournemouth, 1920 (for 1919), pp. 334–5. [p. 218.]

Scott, D. H. and Brebner, G. (1893) On the Secondary Tissues in Certain Monocotyledons. Ann. Bot., vol. VII, 1893, pp. 21–62, 3 plates. [pp. 41, 126.]

Scott, R. (Mrs D. H.) (1909) The Contractile Roots of the Aroid *Sauromatum guttatum*. Rep. Brit. Assoc. Adv. Sci., Dublin, 1909 (for 1908), pp. 910–11. [p. 209.]

Scott, R. (Mrs D. H.) and Sargant, E. (1898) On the Development of *Arum maculatum* from the Seed. Ann. Bot., vol. XII, 1898, pp. 399–414, 1 plate. [p. 209.]

Seifriz, W. (1920) The length of the life cycle of a climbing bamboo. Amer. Journ. Bot., vol. VII, 1920, pp. 83–94, 5 text-figs. [p. 207.]

Seifriz, W. (1923) Observations on the Causes of Gregarious Flowering in Plants. Amer. Journ. Bot., vol. X, 1923, pp. 93–112, 1 plate. [p. 206.]

Serguéeff, M. (1907)   Contribution à la morphologie et la biologie des Aponogétonacées. Université de Genève. Thèse...Docteur ès sciences, Institut de Botanique. Prof. Dr. Chodat, 7me Sér., VIIIme fasc., 1907, 132 pp., 5 plates, 78 text-figs.
[p. 71, and Fig. xlvi, p. 69.]

Solereder, H. (1913)   Systematisch-anatomische Untersuchung des Blattes der Hydrocharitaceen. Beihefte zum Bot. Centralbl., vol. XXX, Abth. I, 1913, pp. 24–104, 53 text-figs.
[p. 125.]

Solms-Laubach, H.,   Weizen und Tulpe. II. Die Geschichte der Tulpen, pp. 37–116,
Graf zu (1899)   1 plate. Leipsic, 1899.
[p. 209.]

Spoehr, H. A.   See MacDougal, D. T., Richards, H. M., and Spoehr, H. A. (1919).

Stapf, O. (1904)   On the Fruit of *Melocanna bambusoides*, Trin., an Endospermless, Viviparous Genus of Bambuseae. Trans. Linn. Soc. Lond., Bot., Ser. II, vol. VI (1901–5), pt. ix, 1904, pp. 401–25, 3 plates.
[pp. 48, 200, 206, 212.]

Sterkx, R. (1900)   Recherches anatomiques sur l'embryon et les plantules dans la famille des Renonculacées. Mém. de la Soc. Roy. des Sciences de Liége. Ser. III, vol. II, No. 2, 1900, pp. 1–112, 24 plates.
[Fig. cxxxvi, p. 171.]

Theophrastus (Hort)   Enquiry into Plants, with an English translation by Sir Arthur
(1916)   Hort, 2 vols. London, 1916.
[p. 148.]

Tieghem, P. van   Sur les feuilles assimilatrices et l'inflorescence des *Danae, Ruscus*
(1884)   et *Semele*. Bull. de la Soc. Bot. de France, vol. XXXI (Ser. II, vol. VI), 1884, pp. 81–90.
[pp. 146, 148.]

Tylor, E. B. (1890)   The Winged Figures of the Assyrian and other Ancient Monuments. Proc. Soc. Bibl. Arch., vol. XII, 1890, pp. 383-93, 4 plates.
[p. 192.]

Vavilov, N. I. (1922)   The Law of Homologous Series in Variation. Journ. of Genetics, vol. XII, 1922, pp. 47–89, 2 plates.
[pp. 226, 227.]

Velenovský, J. (1907)   Vergleichende Morphologie der Pflanzen, vol. II. Prague, 1907.
[pp. 33, 43, 50, 51, 57, 94.]

Wheldale, M.   See Onslow, M. Wheldale.

White, J. H. (1907)   On Polystely in Roots of Orchidaceae. University of Toronto Studies. Biological Series, No. 6, 1907, 20 pp., 5 plates.
[p. 28.]

Willis, J. C. (1922)   Age and Area. Cambridge, 1922.
[pp. 6, 202.]

Willis, J. C. (1923)   The Origin of Species by Large, rather than by Gradual, Change, and by Guppy's Method of Differentiation. Ann. Bot., vol. XXXVII, 1923, pp. 605–28.
[pp. 210, 212.]

Wittmack, L. (1896)   Die Keimung der Cocosnuss. Ber. d. deutsch. Bot. Gesellsch., vol. XIV, 1896, pp. 145–50, 2 text-figs.
[p. 160.]

Woffenden, L. M.   See Priestley, J. H. and Woffenden, L. M. (1922).

Woodhead, T. W.   Notes on the Bluebell. The Naturalist, No. 566, 1904, pp. 81–8,
(1904)   1 plate, 12 text-figs.
[pp. 18, 21, and Fig. vi, p. 18.]

Woodward, A. Smith (**1923**)   Presid. Add. to Linn. Soc., Proc. Linn. Soc., 135th Session, 1923, pp. 27–34, 1 text-fig. [p. 230 *et seq.*]

Worsdell, W. C. (**1908**)   A Study of the Vascular System in certain Orders of the Ranales. Ann. Bot., vol. XXII, 1908, pp. 651–82, 2 plates, 4 text-figs. [p. 112.]

Yapp, R. H. (**1908**)   Sketches of Vegetation at Home and Abroad. IV.—Wicken Fen New Phyt., vol. VII, 1908, pp. 61–81, 1 plate, 7 text-figs. [Fig. xxvi, p. 49.]

INDEX

[Names of authors are not included, since page references are given in connexion with the titles in the bibliography, which thus serves as an index of authors' names. Names of families are not indexed when the reference is merely incidental.]

250

INDEX

INDEX

CAMBRIDGE: PRINTED BY W. LEWIS AT THE UNIVERSITY PRESS

Printed in the United States
By Bookmasters